Dr Kesorn Pechrach Weaver

PIEZOELECTRICS

IN

PROSTHETICS

Energy Harvesting

Piezoelectrics
in
Prosthetics
Energy Harvesting

Dr Kesorn Pechrach Weaver

Pechrach Publishing

Piezoelectrics in Prosthetics: Energy Harvesting

By
Dr Kesorn Pechrach Weaver

ISBN 978-0-9931178-2-4

PECHRACH PUBLISHING
7 Boundary Road, Bishops Stortford, Hertfordshire, CM23 5LE, England, United Kingdom. Tel: (+44) 1279 508020, +44(0) 7779913907 and +44(0)7443426937

Published 2016 by Pechrach Publishing

Every reasonable attempt has been made to identify owners of copyright. Errors or omissions will be corrected in subsequent edition. Although the authors and publisher have made every effort to ensure that the information in this book was correct at the time of going to press, the authors and publisher do not assume and hereby disclaim any liability to any party for any loss, damage, or disruption caused by errors or omissions.

This book is dedicated to people who work
in Prosthetics

Acknowledgments

I would like to thank my beautiful son, Neran J. P. Weaver for his non-stop supporting.

Many Thanks to my family in Thailand and my family in Britain for always believe in me.

Thanks to Prof. J.W. McBride, Prof. J. Swingler, Prof. Suleiman M Sharkh, Peter Wheeler and Peter Wilkes for their support.

I would like to thank and Dr Krisna R. Torrissen for a big encouraging and enormous support, to push me forwards few more steps.

A special thanks to Dr P M Weaver for looking after our family while I write this book.

I would like to thanks my project colleagues: Prof. P. Manooonpong, Dr N. Hatti1, Dr K. Komoljindakul, and Dr K. Tungpimolrut and J. Phontip. We have done a great job and enjoy working together. I would make it real, bring it forward and make it successful.

I would like to say special thank you to K. Carpenter, CEO of Rehabilitation Research Institute (RRI), Seattle, WA, USA for donate the prosthetic legs for our research project.

Thanks to Khon Khan University, King Mongkut's University of Technology Thonburi, University of Southampton, ATPER and OSTC Brussels for providing up to date research information in Europe and funding sources.

My best friends Rajiya Sultana, Anna Wlodarczyk and Kanae Jinkerson for their cheerfully support.

Finally, I would like to thanks the professional, specialists, medical staffs in Rosie hospitals, Addenbrook hospitals, Portsmouth hospitals and Parsonage Lane Surgery.

Introduction

The first time when I got to know and physically touch a piezoelectric when I was working in the laboratory at University of Southampton, UK. The small piece of this material was so thin and amazing me.

I had been living, eating and sleeping with these piezoelectrics while I used them in my experiments. That time I used it as an actuator to trip the circuit breakers when the 22 kA short circuit current was fed through the main feeders.

That when I started to realise this piezoelectric not only need supply power in order to make it work as a switching devices but it also can generate electricity out from any small vibration.

Then, I used the piezoelectric to harvest a lot of energy from many types of research fields and one of them is piezoelectric energy harvesting in prosthetics.

I hope this book will help you to more understand the piezoelectric's behaviour and their applications.

Kesorn Pechrach Weaver

25 April 2016

England, UK

Table of Contents

Acknowledgement

Table of Figures

the prosthetic foot

voltage of MFC

List of Tables

Nomenclature

P   Force

M   Mass

D   Damper

K   Spring

y   Displacement

v   Velocity

is   Current

C   Capacitance

Φ   Electric field

Vre   Voltage

1/R   Reciprocal of Resistance

1/L   Reciprocal of Inductance

Vp   Voltage waveform

Rp   Output resistance

Cp   Output capacitance

ξ   Damping factor

S1 and S2   Complex frequency

Rinl   Resistance for coupling

SW1, SW2   Switches

Rre   Resistance for resonance

Rp   Internal resistance

Lre   Inductance for resonance

Cin   Capacitance for coupling energy

Cre   Capacitance for resonance

Fo   Resonance frequency

MPF   Macro Fiber Composite

MPP   Maximum Power Point

RL   External load resistor

Ri   Internal resistor

ZIn_Total   The relation between impedance from
            resistance

CHAPTER 1

ENERGY HARVESTING

This works present the design of an energy harvesting system using smart materials for self power generation of upper and lower prosthetic legs. The smart materials like Piezo Composites, Piezo Flexible Film, Macro Fibre Composites, and PZT have been employed and modified to be appropriately embedded in the prosthesis.

The movements of the prosthesis would extract and transfer energy directly from the piezoelectric via a converter to a power management system. Afterward, the power management system manages and accumulates the generated electrical energy to be sufficient for later powering electronic components of the prosthesis.

The experimental results of energy harvesting and efficiency in peak piezoelectric voltages during step up and continuous walking for a period of time. One of the most interesting sources for energy harvesting is environmental vibrations.

The devices that have been used are piezoelectric, electromagnetic, electrostatic, pyroelectric, photovoltaic and thermoelectric. The conversion of harvesting energy is very good in scalability, capability, high energy density and compatible with standard electronic technology. In addition, the piezoelectric could be coupled to a mechanism to perform opening of the contacts in the switching devices [1-3].

The green energy harvesting here will emphasis use of piezoelectric devices in Prosthetic legs. Geometric parameters, beam, mass and resistive electric loads significantly influence the output power [4].

The piezoelectric CMOS harvesting could bypass the input voltage and recover some energy to increase the energy during negative piezoelectric voltage [5].

1.1 Piezoelectric PZT

The smart material used for energy harvesting is piezoelectric PZT (Lead Zirconate Titanate) as shown in Figure 1.

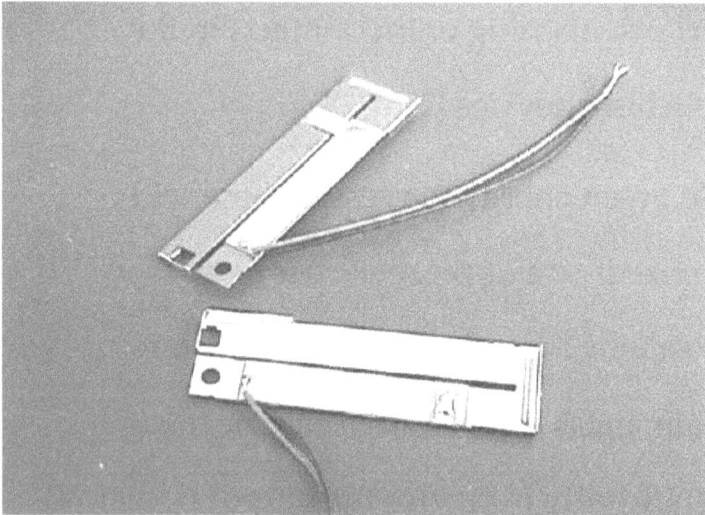

Figure 1.1: Piezoelectric PZT [1-3, 10]

This piezoelectric PZT comprises two arms formed from a single metal substrate and two piezo-ceramic plates [1-3]. The first arm of the piezoelectric produces a downward movement which provides an angle for the movement amplification of the second arm as shown in Figure 2.

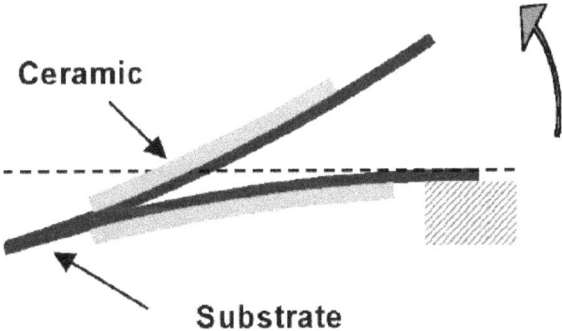

Figure 1.2: Diagram of the PZT piezoelectric [1, 10]

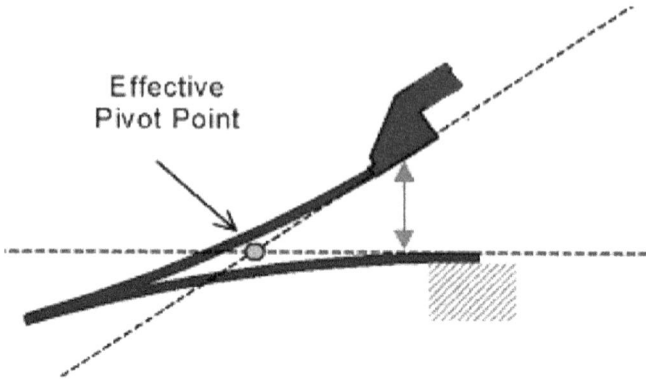

Figure 1.3: The effective pivot point [2]

The advantage of the planar bimorph structure is that it forms an effective pivot point approximately half way along the length of the Piezoelectric [6] as shown in Figure 3. This would help to produce significant amount of power generated from piezoelectric.

Piezoelectrics in Prosthetics: Energy Harvesting

1.2 The Macro Fibre Composite (MFC)

The MFC consists of rectangular piezo ceramic rods sandwiched between layers of adhesive and electrode polyimide film. This film contains interdigitated electrodes that transfer the applied voltage directly to and from the ribbon shaped rods [7-9]. As a thin, surface conformable sheet it can be applied, normally soldered to various types of structures or embedded in a composite structure.

1.3 MFC Piezoceramic Structure

The interdigitated electrode pattern on polyimide film on both and bottom permits in plane polling of piezoceramic d33 (elongator mode) versus d31 (contractor mode) advantage. In addition, the structure epoxy inhibits crack propagation in ceramic bonds components together.

1.4 Energy Harvesting Circuit

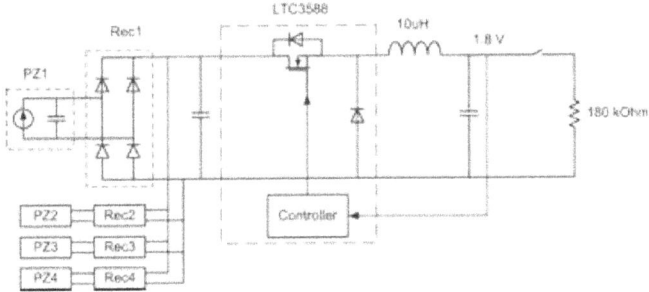

Figure 1.4: PZT Energy harvesting circuit diagram [10]

The circuit diagram for converting kinetic energy from the movement of the prosthetic legs and foot prosthesis to electrical energy is shown in Figure 4 and 5, respectively.

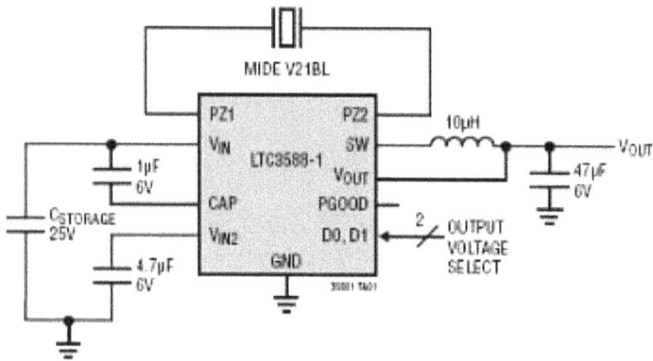

100mA Piezoelectric Energy Harvesting Power Supply

Figure 1.5: MFC Energy harvesting circuit diagram [10]

The movement of the prosthetic legs and the foot prosthesis would generate energy directly from PZT and MFC piezoelectric. The integrated low loss full wave bridge rectifier with a high efficiency buck converter was designed to connect directly to PZT and MFC piezoelectric.

CHAPTER 2

PIEZOELECTRIC

The development of piezoelectric energy harvesting is a new technology for electrical protection systems. A range of alternatives have been investigated such as the redesign of the normal mechanism to permit smaller magnets.

2.1 Planar Bimorph Energy harvesting

The energy harvesting used for the present investigation in this book was a planar bimorph energy harvesting which D31 benders give better motion. To increase the motion and force, a rectangular energy harvesting form to increase the anisotropy of the beam and a cantilever system. It gains a motion in order to interact with a mechanism. This is because a cantilever has many times the deflection of a simply supported beam of the same section and length [11, 12]. The energy harvesting employs a novel geometry, which produces very large movement from a simple and compact structure [11].

2.2 Planar Bimorph Operation

The energy harvesting comprises two arms formed from a single metal substrate and two piezoelectric ceramic plates. The operating principle is shown in Figure 2.1.

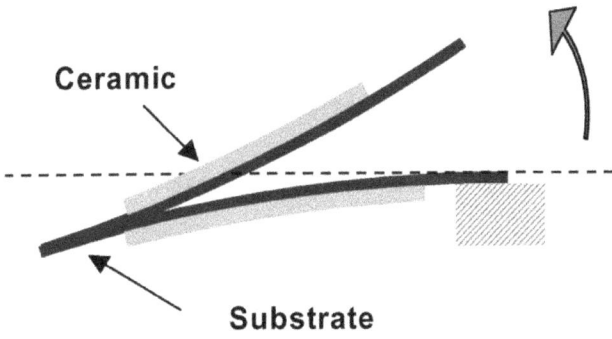

Figure 2.1: Schematic diagram of the operation of the planar bimorph energy harvesting [1, 10, 13, 14]

The planar disposition of the two ceramic plates means that the high fields can be applied without danger of repolarising the ceramic. This is different from a conventional bimorph structure.

The first arm of the energy harvesting produces a downward movement which provides an angle for the movement amplification of the second arm. The total movement is slightly less than could be obtained from a single beam of twice the length.

However, the stiffness of a single beam has cubic relationship to the length. So that only very low forces could be obtained this way.

2.3 Planar Bimorph Structure

A key advantage of the planar bimorph structure is that it forms an effective pivot point approximately half way along the length of the energy harvesting as shown in Figure 2.2.

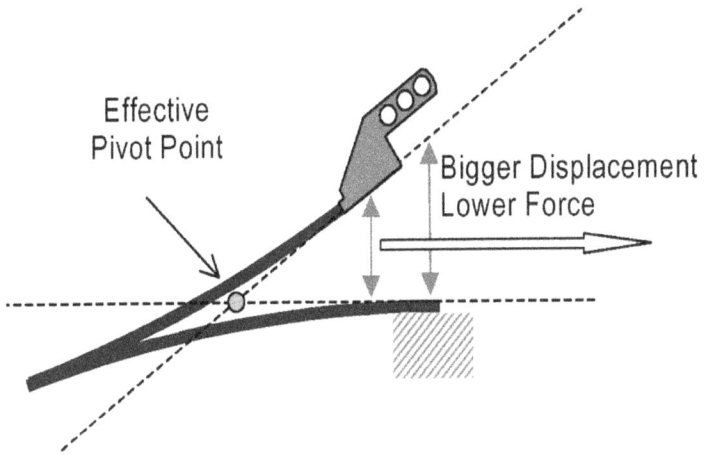

Figure 2.2: Effective pivot in planar bimorph energy harvesting [1,10, 13, 14]

Therefore, a high displacement can be obtained with useful force output. This results in a much more efficient energy harvesting than a conventional single beam design.

2.4 Surface Structure

The piezoelectric ceramic is a commercial grade lead zirconate titanate. It is selected for its high coefficient 330 pm V to achieve a large displacement in a bending energy harvesting. The formed and compacted ceramic is sintered and then machined to size. The details of electro-ceramic processing have shown in [33]. This forms a dense grain structure with typical grain sizes of 15- m diameter, as shown in the surface profile of Figure 3.4. A metallic electrode is then applied to the ceramic. The ceramic plates are bonded to the metal substrate and electrical connections made to complete the assembled energy harvesting.

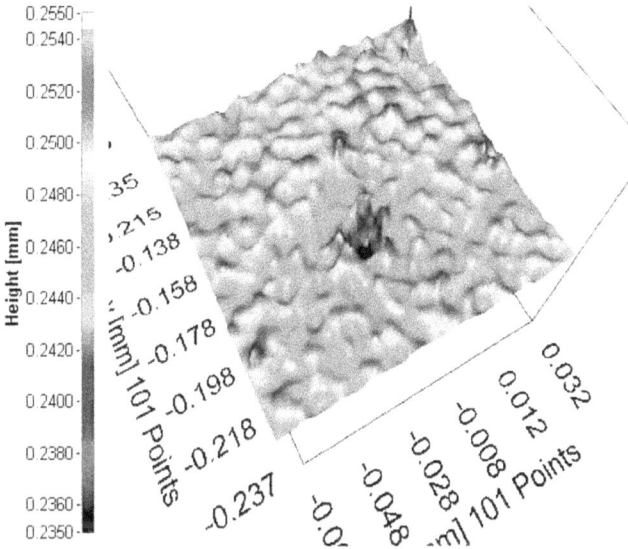

Figure 2.3: Surface structure of piezoelectric ceramic [1-3, 10]

2.5 Piezoelectric and Prosthetics

When applied in prosthetic mechanism the energy harvesting behaviour can be approximated by a simple mass spring system [1, 13, 14]. The spring rate is independent of the applied voltage, but the unloaded position varies with voltage.

However, the complex thermal behaviour of the piezoceramic materials can often severely limit the performance of the system. This leads to increased cost, complexity and reliability issues. Not only do the basic piezoelectric parameters vary with temperature, but also it is dominant.

The effect is the rapid increase in the hysteresis as the temperature reduces. This is especially true for the very active soft ceramics used to provide maximum performance from the energy harvesting. Moreover, the coercive field vary with temperature. The relationship between position and applied charge or voltage is also temperature dependent.

It is possible to implement a control system that applies a temperature variable reverse field and charge rate using a low cost microcontroller. This system gives substantially improved performance at low temperatures. In fact the energy harvesting force and deflection increases as the temperature decreases.

This improves tolerance to variations in position of both the energy harvesting and the mechanism being actuated. This can lead to cost benefits in the deployed mechanism through the use of lower precision manufacturing and assembly techniques.

2.6 Piezoelectric Film

Loreto Mateu and Francesc Moll [15] used a piezoelectric film inserted inside a shoe, to connect the load only when there is enough energy stored in the capacitor. piezoelectric material is a PVDF film. These films are only metalized in the plane perpendicular to direction 3, so that D1 = D2 = 0.

The piezoelectric element is placed inside the insole of a shoe. The mechanical walking activity cause a mechanical deflection on the piezoelectric films.

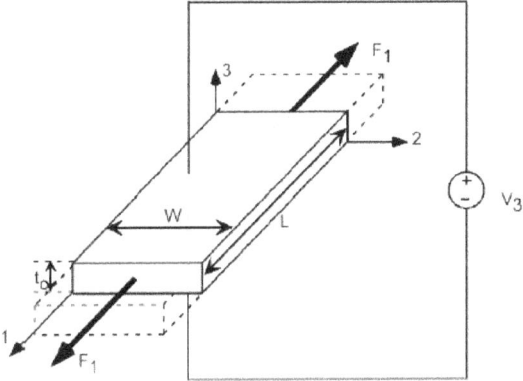

Figure 2.4: Mechanical diagram

The phenomenon of piezoelectricity is described by the piezoelectric constitutive equations:

$$S_i = s_{ij}^E T_j + d_{li} E_l$$
$$D_m = \varepsilon_{mn}^T E_n + d_{mk} T_k$$
$$\text{for } i, j, k = 1, \ldots, 6 \text{ and } l, m, n = 1, 2, 3$$

where T is the applied mechanical stress [N/m2], E is the applied electric field [N/C], d corresponds to the piezo strain [(C/m2)/(N/m2)], "T is the permittivity [F/m] under conditions of constant stress, D is the electric displacement [C/m2], S is the mechanical strain [m/m], and sE is the compliance tensor [m2/N] under conditions of constant electrical field.

Nordic nRF24E1 microcontroller was used with an integrated RF transmitter at 1 Mbps [8]. This device needs a 3 ms phase of configuration time with a power consumption of 5.7 mW, then a short interval (0.3 ms approximately) to access the ADC and RF transmitter, with a power consumption of 27 mW.

In an energy harvesting application, it is desirable to have energy neutral operation [16], that means, to ensure that the energy harvesting transducer is generating at least the same amount of energy as the energy consumed by the electric load.

CHAPTER 3

PROSTHETICS

Prosthetic technology is advancing rapidly devices. what amputees need and what they get can often be quite different.

Socket

This is interfaced the residuum and the prosthesis. It needs to be flexible and have a perfect fit to prevent painful pressure.

Shank

This supports the knee-joint and the foot.

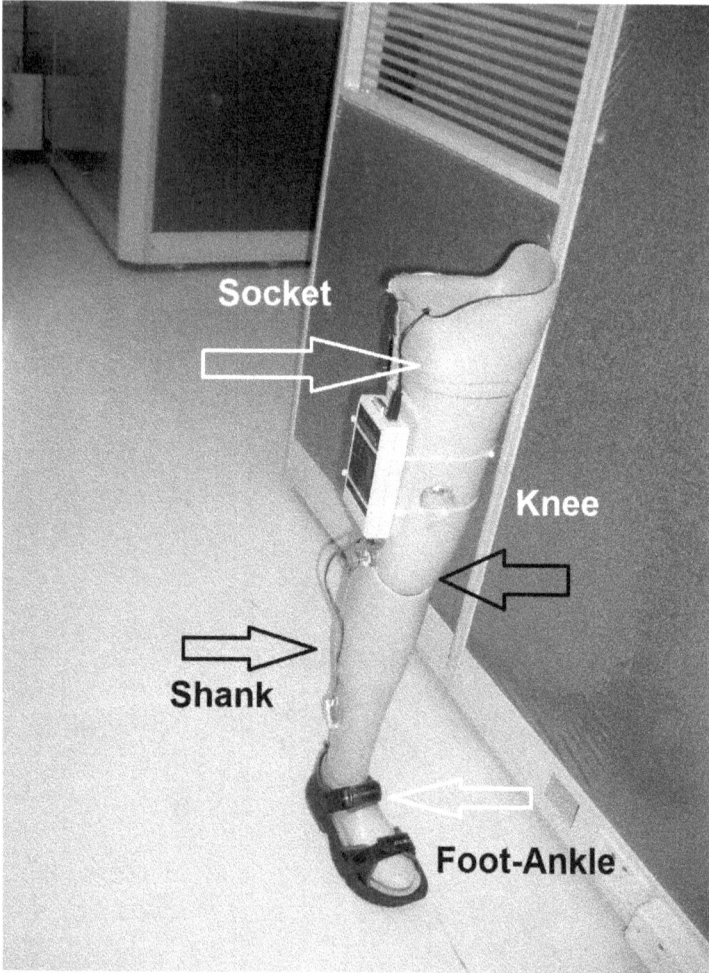

Figure 3.1: Prosthetic leg components

Piezoelectrics in Prosthetics: Energy Harvesting

Foot

It uses to contact to the ground and provides shock absorption and stability.

Knee

There needs a complex connector allowing flexion during swing-phase.

3.1 Prosthetic Leg with harvesting equipments

The period of standing phase the direction of the reaction force exerted by the ground coincides with the direction of the relative motion of the ankle joint.

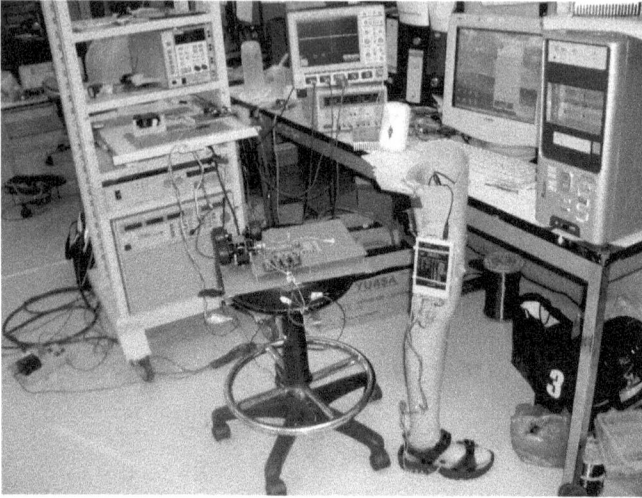

Figure 3.2: Side view of harvesting prosthetic leg in standing position

The power variation in the standing phase can be calculated by considering a model. [17]

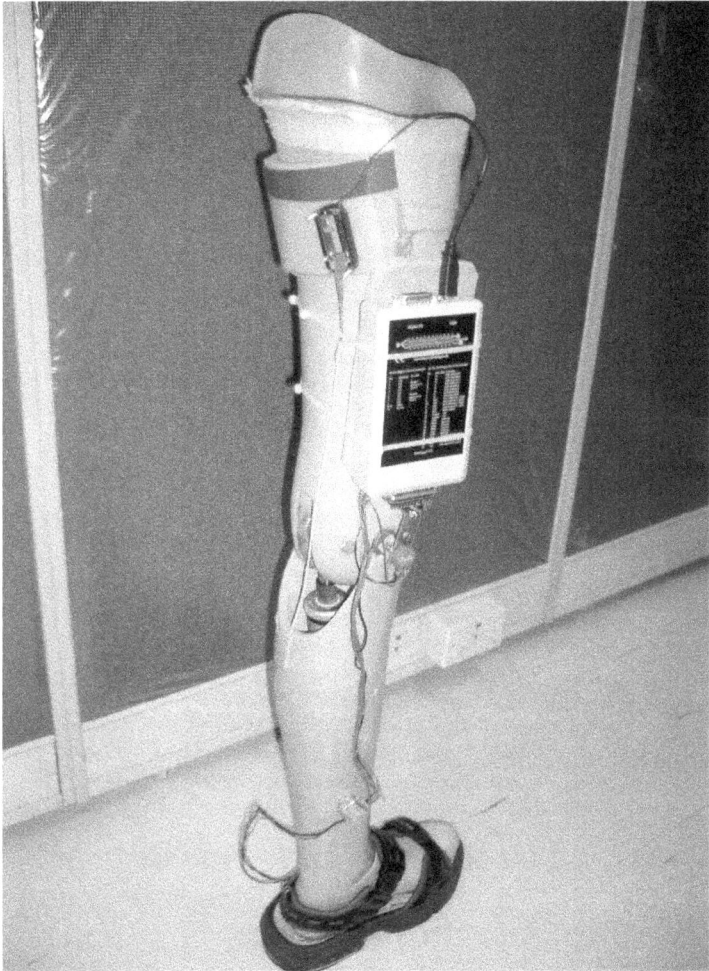

Figure 3.3: Back view of harvesting

prosthetic leg in standing position

3.2 Upper Prosthetic Leg with harvesting equipments

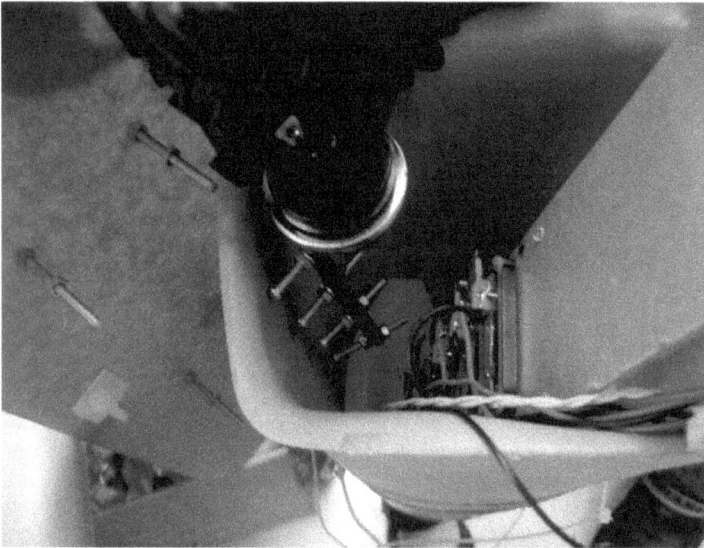

Figure 3.4: Piezoelectric harvesting installed inside the upper prosthetic leg

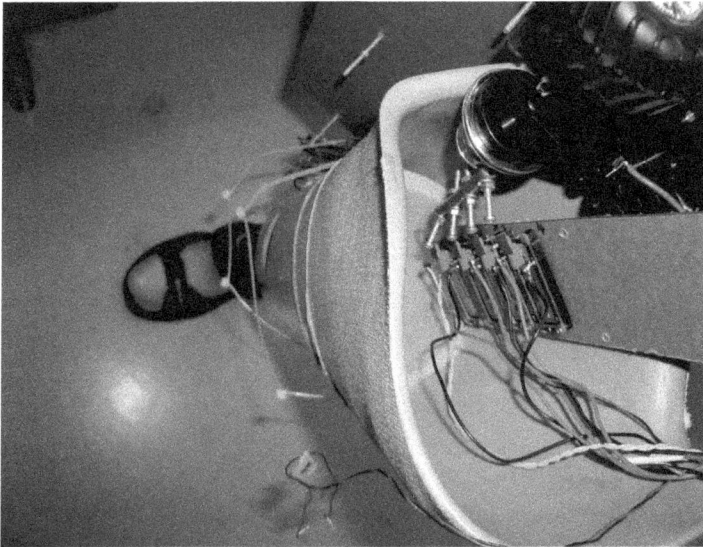

Figure 3.5: Top view of Piezoelectric harvesting installed inside the upper prosthetic leg

There is a plan to embed some piezoelectrics inside the socket area including the hitting harvesting kit. The complete set would be moulded as part of the material for the upper prosthetic leg.

Figure 3.6: Piezoelectric harvesting with hitting mechanism

The harvesting circuit for using with the hitting mechanism consists of hitting hands, cores and low energy consumption motor. This test kit can have a variation speed to find the maximum delay time to create the maximum vibration in the PZT blade.

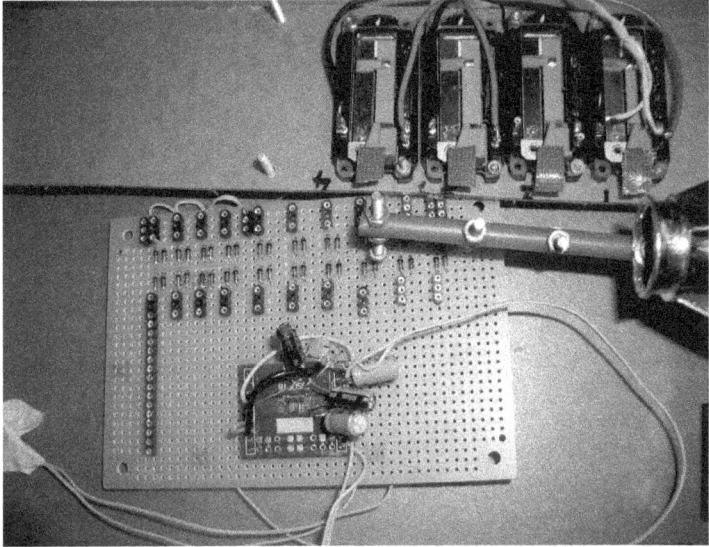

Figure 3.7: Details of harvesting circuit and mechanism

Four sets of PZT piezoelectric connects as series to produce more power from harvesting energy. Four capacitor banks are

used as storage energy tanks for power
which produce from PZT piezoelectric.

3.3 Prosthetic Knee & harvesting equipments

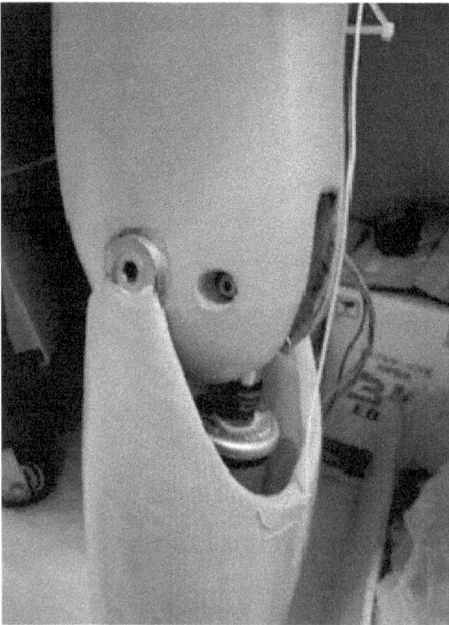

Figure 3.8: The connection inside the
Prosthetic knee

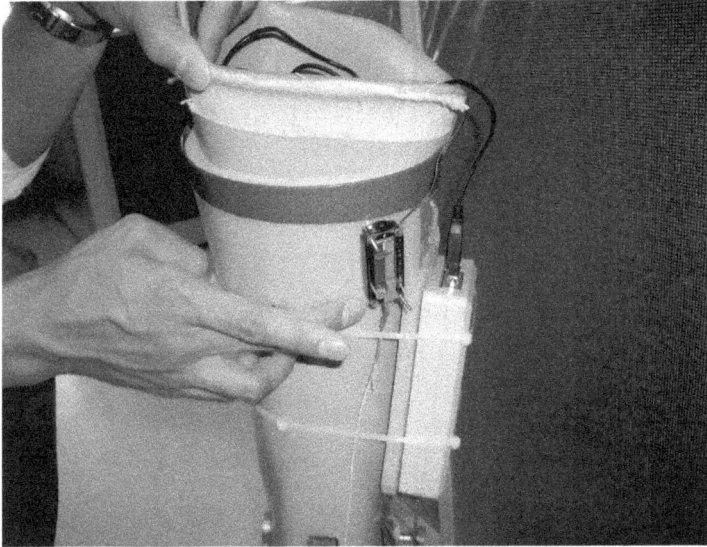

Figure 3.9: Piezoelectric harvesting from the prosthetic knee action

The PZT piezoelectric energy harvesting connect with cables to vibrate the piezoelectric. The cable connect from piezoelectric and attach to the knee mechanism. The cable will activate the PZT piezoelectric to start harvest energy when the prosthetic knee in the walking position while the knee is in the angle or bend position.

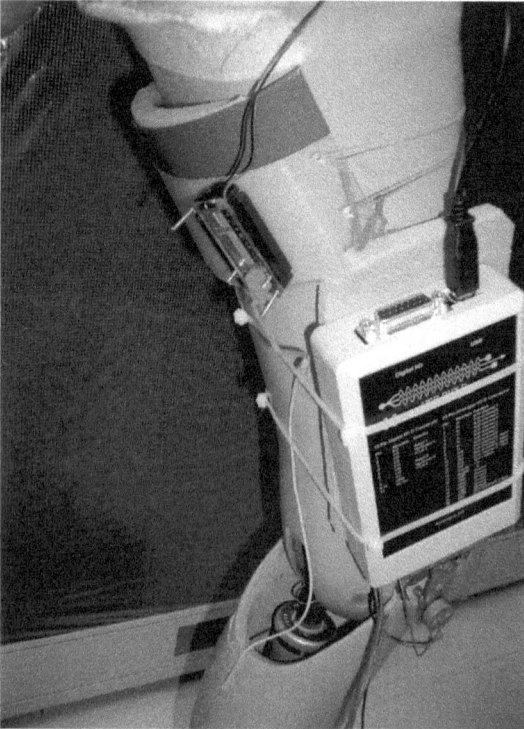

Figure 3.10: Piezoelectric harvesting in the pull position

When the prosthetic knee in the bend position, in the swing back walking or step up or down, the piesoelectric will be in the pull position until the knee position goes back in the status straight position.

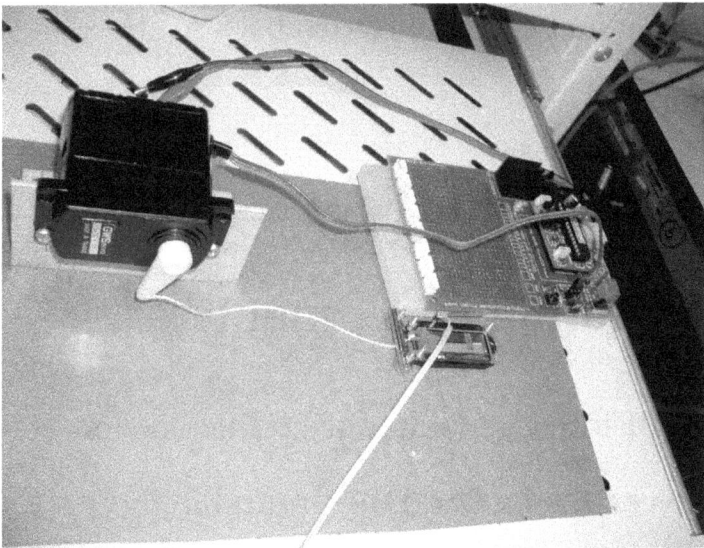

Figure 3.11: Laboratory test with the pull position

The step motor was used to run the dry test to pull the piezoelectric. This simulation in the case of walking forwards, backward, step up and step down which the prosthetic knee would be changed from the straight position to bend and return back to straight again.

3.4 Prosthetic Foot & harvesting equipments

Propulsion is generated from the reaction between the foot and the ground in the standing phase. To improve the propulsion function of the foot prosthesis and stump system, it must shorten the first period of the standing phase when the ankle power is negative and more quickly release energy in the second period. Decreasing the elastic modulus of the material of the energy storing element could greatly increase the energy stored and energy released. The elastic modulus of the material has only a small effect on the propulsion index, but the stored and released energy changes significantly with it. [17]

Considering the motion and power variation in the ankle joint in the standing phase, during the muscles contract and the direction of reaction force exerted by the ground coincides with the direction of the relative motion. The muscular power during that period is defined to be negative.

During the standing phase, the muscles extend, the reaction force is in the opposite direction to the relative motion, and the power is defined as positive. [49-51]

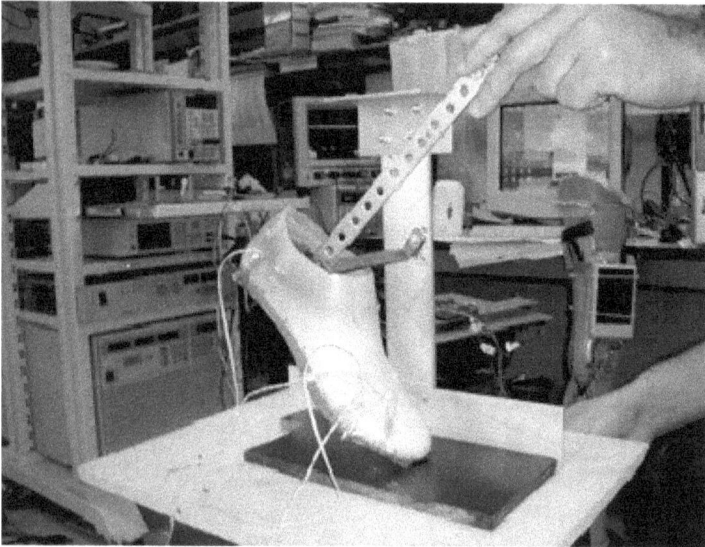

Figure 3.12: Adjustment prosthetic foot position

The test stand can adjust the angle of the prosthetic foot position. To test the suitable positions to harvest energy from foot while walking, standing and swing position.

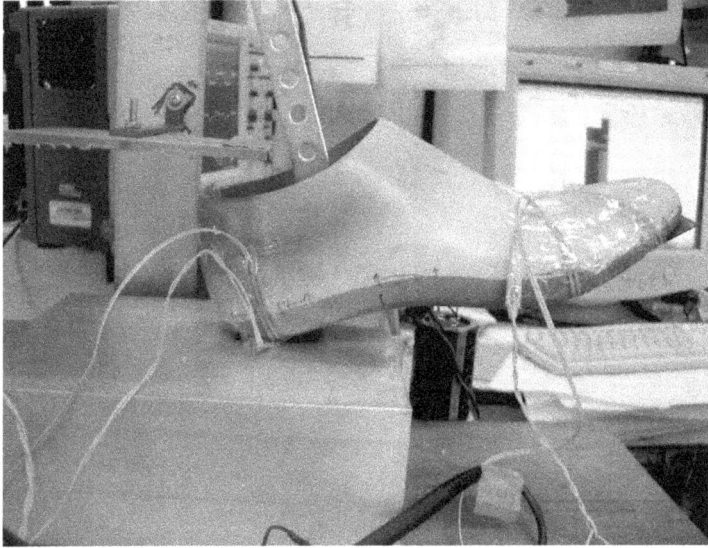

Figure 3.13: Sole of the prosthetic foot

The film piezoelectric were placed in the sole position, in the bottom of the foot. Sole is the first position to touch the ground when walking forward.

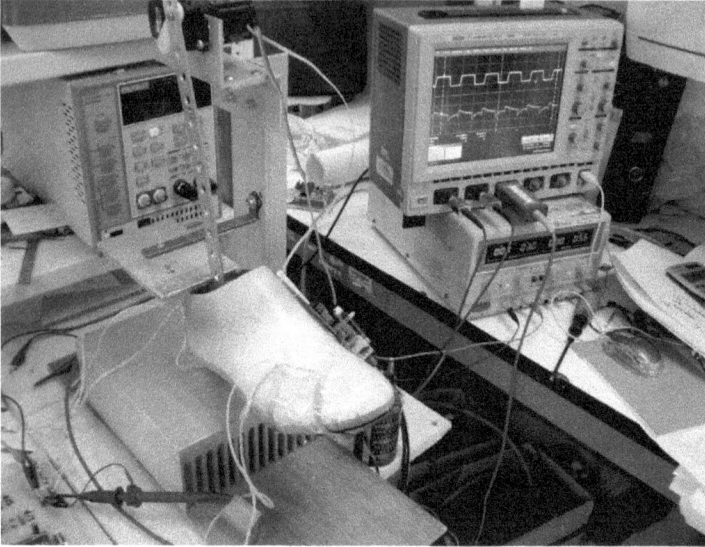

Figure 3.14: Sole harvesting and signals

Figure 3.14 show the piezoelectric energy harvesting at the sole position of the prosthetic foot. The power generation signals from the piezoelectric show on the oscilloscope. When walking, the weight is transferred from the heel to the ball of the foot.

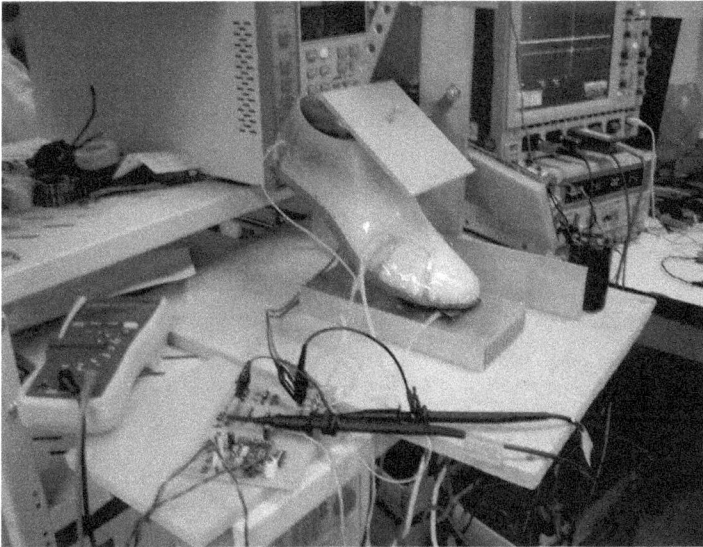

Figure 3.15: Energy harvesting in the ball of the foot

The portion of the foot between arch and toes is called the ball of the foot. This area is the last part to contact with the ground after the sole of foot when walking forwards and the last part to left the ground in the swing state before the foot move forward. However, when the foot moves backward, the sole of foot is the first part to make a contact with the ground.

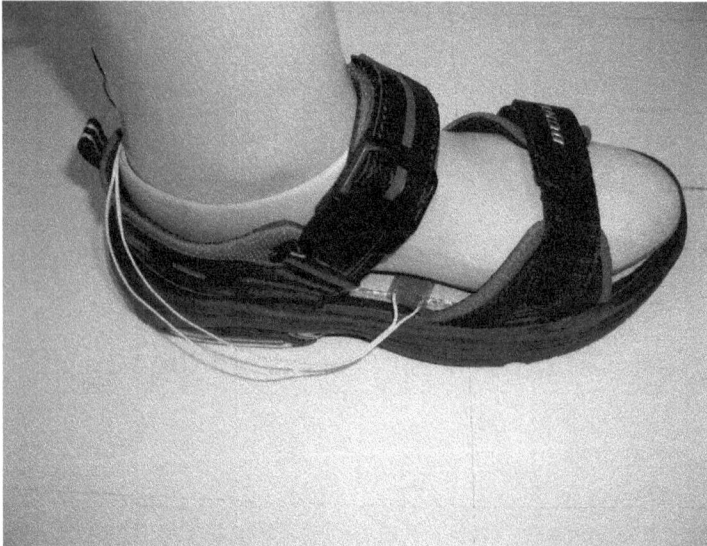

Figure 3.16: Film piezoelectric insert under the prosthetic foot

These film piezoelectrics can be removed to place in different positions. In the beginning, the positions were chose to place at the sole of the foot and the ball of the foot. However, the area between the sole and the ball of the foot can be placed some piezoelectrics to harvest energy by using foot arch support.

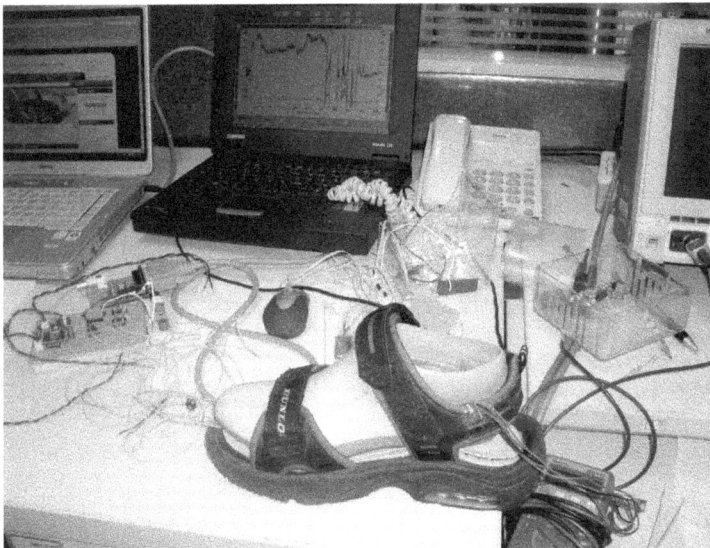

Figure 3.17: Piezoelectric positions under the foot

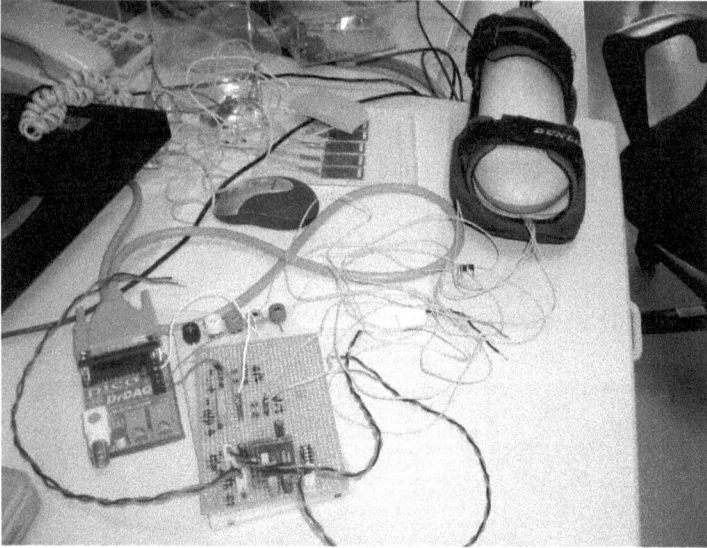

Figure 3.18: Energy harvesting circuit with data acquisition

Arrays of film piezoelectric were planned to underneath the prosthetic foot. The harvesting circuit with capacitor banks were used to install power from harvesting from piezoelectric.

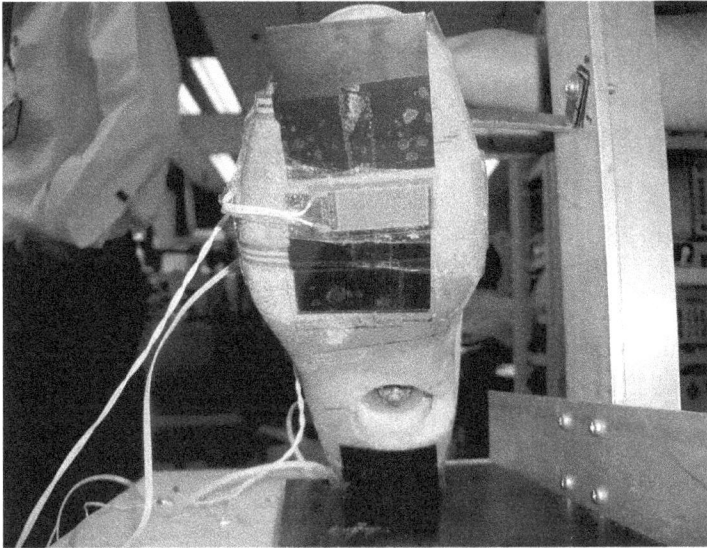

Figure 3.19: Film piezoelectric at the ball of the prosthetic foot

The ball area of the prosthetic leg can place the film piezoelectric to harvest energy more than 10 piezoelectric while at the sole area of the prosthetic foot, there can place more than 6 sets of film piezoelectric.

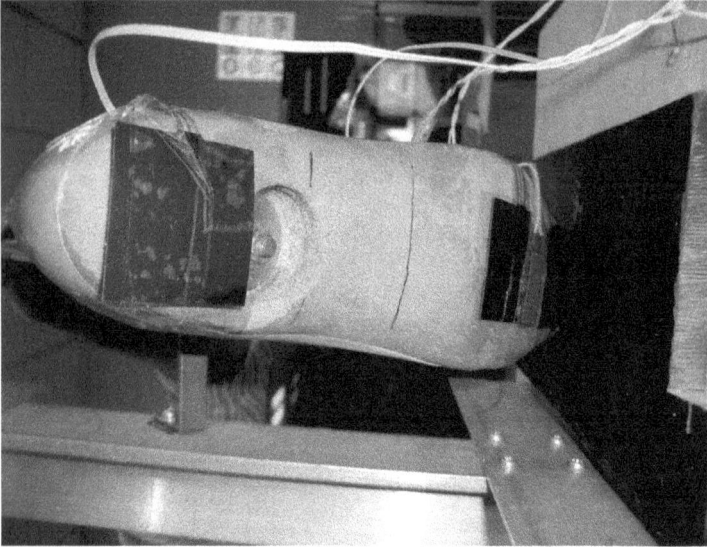

Figure 3.20: Force sensors position

Two sets of force sensors were placed at the ball of the prosthetic foot and one set of force sensor was at the sole of the prosthetic foot. These forces sensors were used to check how much of the weight, the position of the maximum and minimum force in the foot area.

CHAPTER 4

EXPERIMENTAL SETUP

Energy harvesting transform natural energy sources into usable electrical energy, for example, solar energy, thermal energy, wind and vibration energy, etc.

This technology is very attractive for low power electronic devices which include medical devices, smart implants, camera imaging inside the human body and hearing aid devices.

Generating energy from human body vibration is one of the current challenges. In this work, the maximum peak power and peak voltage energy harvesting from PZT piezoelectric generated is much higher than MFC piezoelectric.

Figure 4.1: PZT piezoelectric setup

This figure shows the details of the experimental equipment, four sets of PZT piezoelectric were used to harvesting energy in the upper prosthetic legs.

4.1 Experimental Equipments

The platform used to hold a prosthesis foot as shown in Figure 7.

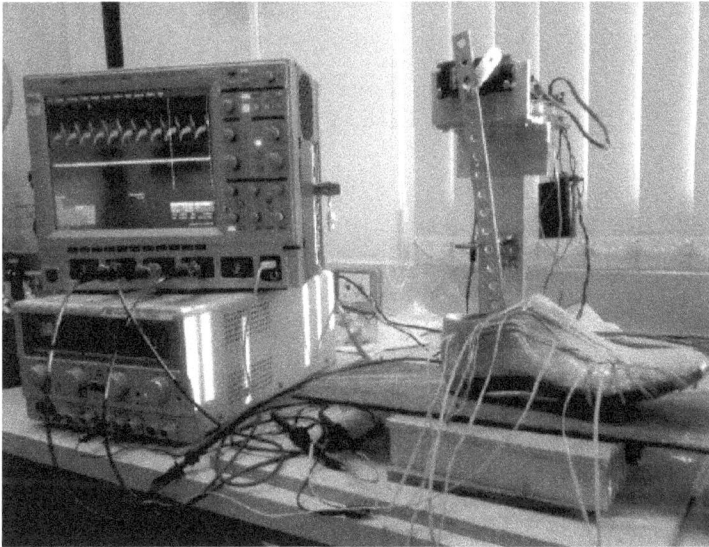

Figure 4.2: Experimental equipments

This can travel in both of the direction of horizontal and vertical. The angle of foot prosthesis is able to set up the moving from 0 degree to 135 degree.

4.2 Installation Position

There are six sets of MFC Piezoelectric were installed underneath the prosthesis foot. This would use to harvesting power out from the lower prosthetic legs.

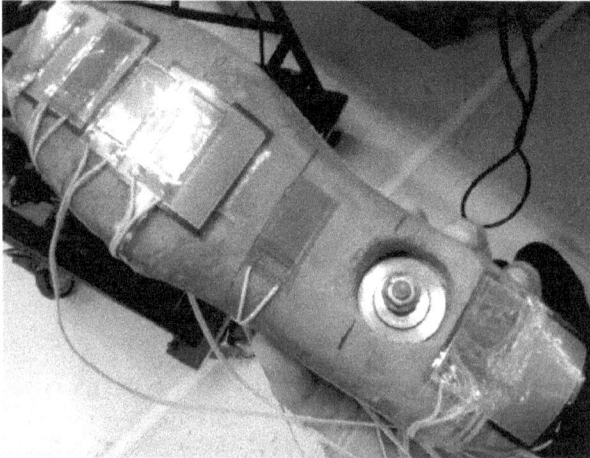

Figure 4.3: MFC installing in the prosthesis foot

Piezoelectrics in Prosthetics: Energy Harvesting

The 1st position of piezoelectric was at the foot sole as shown in Figure 4.3.

4.3 Waveform of PZT

The experimental results from the PZT energy harvesting as shown in the Figure 9, the peak voltage at capacitor is approximately 4 volts and the current out is approximately 2 A.

The peak power is approximately 88.8 µW. which is more than 80% higher than D. Kwon et al [4] but lower than [3] by 76%.

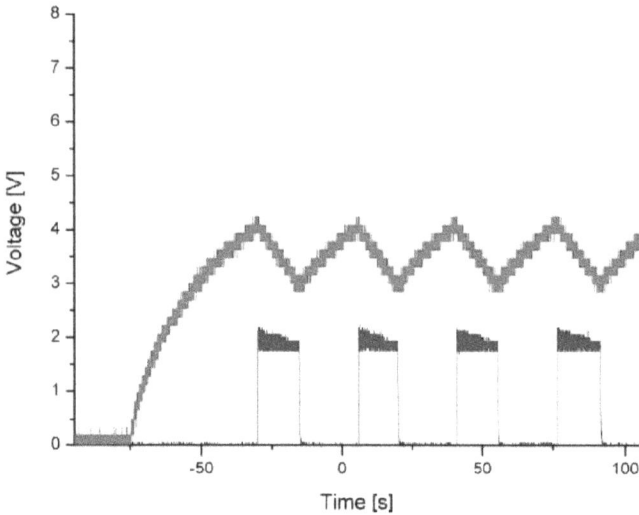

Figure 4.4: The voltage and current of the PZT [10]

4.4 Waveform of MFC

Figure 10 shows the typical waveform of MFC piezoelectric energy harvesting which installed underneath the prosthesis foot.

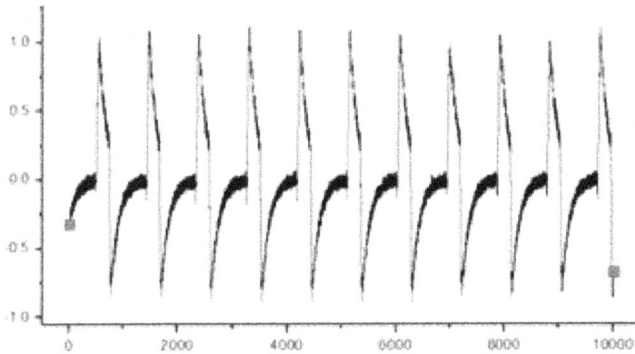

Figure 4.5: Typical waveform of MFC [10]

The maximum voltage is approximately 1 Volts when the foot sole reach the ground. The maximum of the peak voltage, as shown in Figure
11, is about 1.8 Volts and maximum peak power is about 12.5 µW. at the 1st MFC piezoelectric location.

4.5 The Effect of Piezoelectric Position

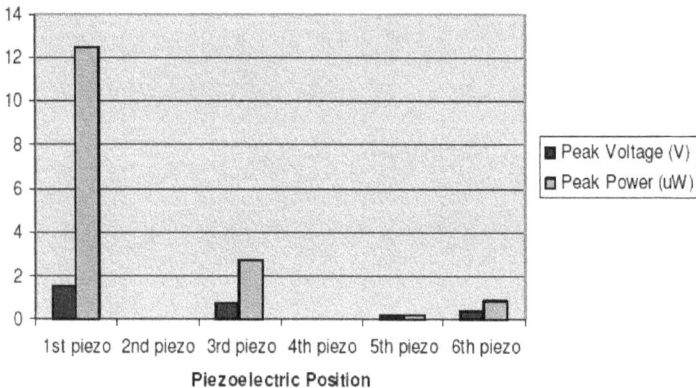

Figure 4.6: Peak voltage and peak power [10]

The first position provides the highest energy harvesting voltage out from the whole foot while the smallest output is on the fourth position. The energy harvesting at the third position is higher than the sixth position but lower than the first position.

4.6 The Effect of Angle Degrees

It shows the highest energy harvesting when the foot sole reach the ground, the maximum of the peak voltage is about 1.2 Volts and the maximum peak power is about 8 μW.

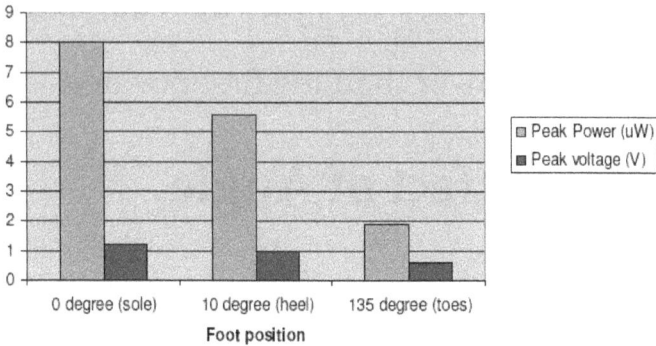

Figure 4.7: Peak voltage and peak power at prosthesis foot positions in degrees [10]

This shows that the upper leg could generate power more than the lower leg at prosthesis foot. This shows the same results as [8]. The degrees of walking [9] also have an effect on the power output. In addition, the prosthesis foot position produces the most energy when the foot sole reach the ground.

Piezoelectrics in Prosthetics: Energy Harvesting

Walter D. Pikey [17] studied the analytical evaluation of an energy-Storing foot prosthesis and found that the power variation of the ankle joint for different cases: a human foot, a foot prosthesis and a conventional foot prosthesis. It was clear showed that the power variation of energy-storing prosthesis is much closer to that of a human foot than that of the conventional foot prosthesis. The values of human feet are greater because muscles generate energy.

Piezoelectrics in Prosthetics: Energy Harvesting

CHAPTER 5

ELECTRO-MECHANICAL ENERGY

This chapter proposes a new electro-mechanical model for piezoelectric ceramic lead zirconate titanate (PZT) employed in energy harvesting systems.

The proposed model consists of a parallel resonant circuit, loss components and discrete switches to emulate behaviours of electrical energy generated from the PZT vibration. To verify the accuracy of the proposed model, its electrical behaviour is experimentally compared with that of the prototype. The results indicate a significant consistency.

Additionally, this paper demonstrates how the proposed model performs in energy harvesting system design process. The results confirmed that the proposed model can be practically utilized.

The main parameters will be determined and modelled. To confirm its accuracy, the electrical behaviours of the proposed model is simulated by using PSPICE program and experimentally verified with the prototype. Moreover, to show the utilization of the proposed model, a buck-converter based harvesting system behaviour is simulated as a test case.

5.1 Energy Characteristic

Currently, this is an active research and development in utilizing alternative energy sources such as wind, solar, geothermal energy, etc. Nonetheless, for applications those require low power have a concept of using surplus vibration energy from, for example, running industrial electric motor or human motions [10].

To convert these mechanical energy to electrical energy, a well-known device so-called "piezoelectric" is employed. Electrical energy from the piezoelectric is, however, still very low; therefore, piezoelectric devices need to be integrated with energy harvesting circuits to achieve effectively officiate energy storage and energy transmission.

The mechanical and electrical behaviours, including the characteristics of the piezoelectric devices have been observed. The energy flow and impedance modelling have been investigated by [18]. However, the performance of Piezoelectric PZT is considered as one of the promising piezoelectric that provide high output power and energy.

The PZT devices are based on cantilever beam structure. In previous research [1-3, 10], a prototype for PZT with cantilever beam was modelled based on a dynamic analysis of the mechanical part and its response in voltage waves in an open circuit test.

It was also tested using load resistance to find its maximum power where the result revealed that an equivalent circuit inside the PZT was a series RLC circuit which have been found in general research articles. In this chapter, the aim is to invent a new PZT model capable of emulating a better electrical behaviours in low frequency range operation.

The behaviours and characteristics of a novel structural piezoelectric ceramic lead zirconate titanate (PZT) shown in Figure 5.1. It comprises two piezoelectric ceramic plates (arms) connected together with a substrate. [1-3, 10]

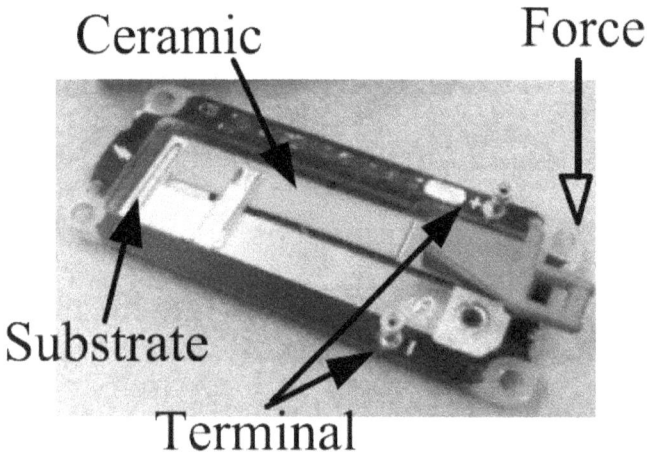

Figure 5.1: Components of the PZT [1-3, 10]

This structure of piezoelectric PZT results in a higher energy conversion efficiency than many ranges of low frequency applications [1-3].

5.2 Electero-Mechanical Behavior

The proposes analysis and modelling techniques to emulate the electrical behaviours for the PZT based on its experimental results.

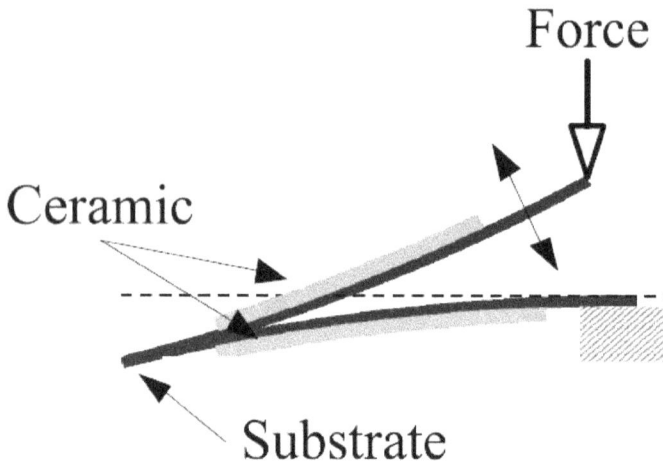

Figure 5.2: A structure of the PZT [1-3, 10]

In order to find parameter which can reveal characteristics of this piezoelectric PZT, it is important to understand the characteristic of the electrical energy which cab be generation from the action of the mechanical system. This production energy is created by the vibration action.

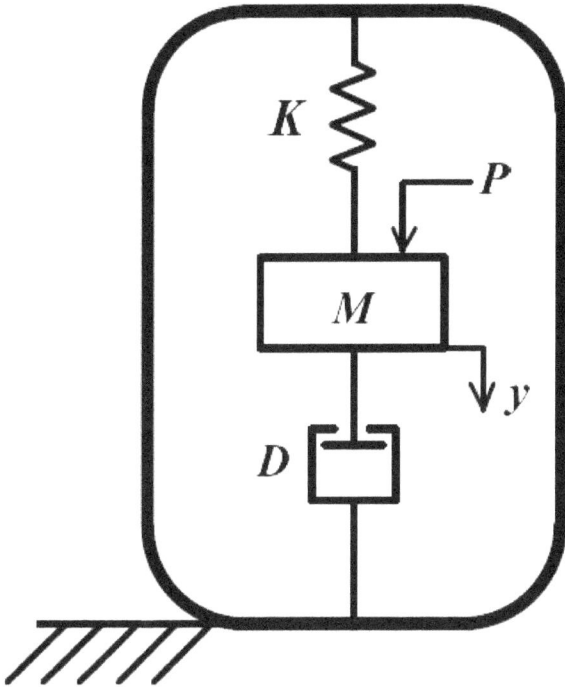

Figure 5.3: Mechanical system [20,21]

The modelling starts with the analyse both the mechanical system and the electrical systems as figure 5.3 and 5.4.

Figure 5.3 shows a mechanical system which consists of Force (P), a mass (M), a damper (D), a spring (K), and displacement (y) [19].

This mechanical system can be used to represent the natural respond of the piezoelectric devices and is expressed as [1-3, 10].

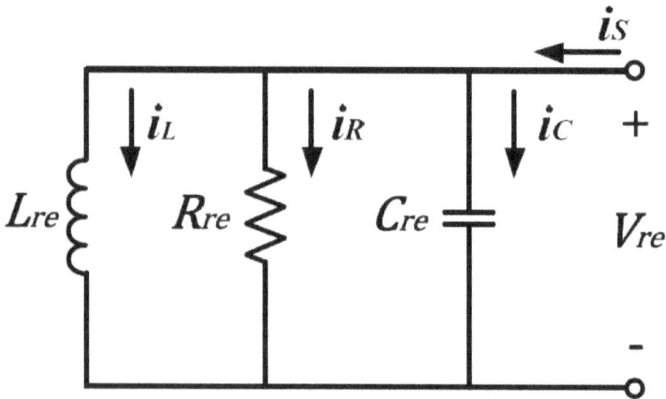

Figure 5. 4: Electrical system [20,21]

In order to emulate the mechanical dynamic response of the proposed piezoelectric PZT. Therefore, in the electrical system as shown in figure 5.4, a parallel resonance circuit could be used as following:

$$M\frac{dy^2}{dt} + D\frac{dy}{dt} + Ky = P$$

$$C\frac{d\varphi^2}{dt} + \frac{1}{R}\frac{d\varphi}{dt} + \frac{1}{L}\varphi = i_S$$

From these relationship the mechanical components can transform as an equivalent circuit to electrical component parts are

shown in Table 5.1.

Mechanical Part	Electrical Part
Force: P	Current: is
Mass: M	Capacitance: C
Displacement: y	Electric field: φ
Velocity: $\dot{y}$	Voltage: Vre
Damper: D	Reciprocal of Resistance: $\frac{1}{R}$
Spring Constant: K	Reciprocal of Inductance: $\frac{1}{L}$

Table 5.1: Mechanical and Electrical

parameters [20,21]

CHAPTER 6

ELECTRO-MECHANICAL MODELLING

This chapter presents the electro-mechanical modelling technique for the piezoelectric PZT (Lead Zirconate Titanate). The electromechanical characteristic of the studied PZT would be analyzed based on experiments while the investigation of the energy generation was shown in Figure 6.1 which used a DC motor which employed as the mechanical force source. It used to control the action which would make contacts to one of the piezoelectric PZT arm at the frequency 1 Hz. The output voltage waveform, *VP*, generated from the PZT is observed, investigated and measured the value at each touching step.

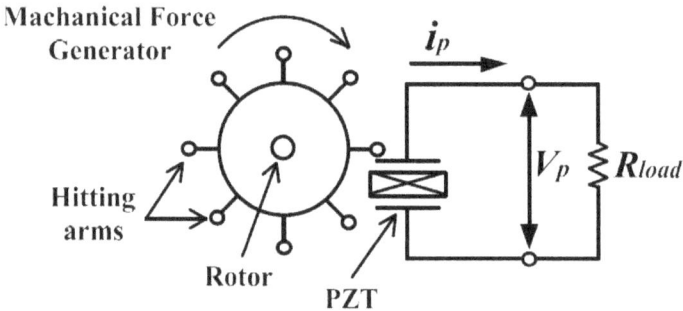

Figure 6.1: Experimental setup with hitting force [20,21]

The electrical energy generated by PZT can be either in the form of voltage source or current source. The response of the PZT would be emulated by these techniques.

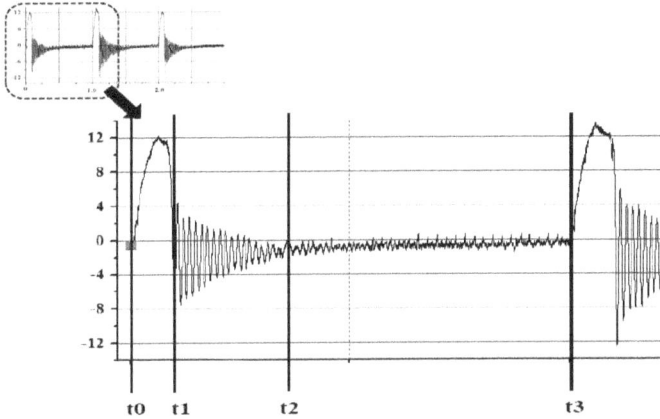

Figure 6.2: Open-circuit voltage waveform of the PZT prototype [20,21]

The result revealed that there were two response intervals which generated voltage waveforms caused by

(A) between period t0 and t1

The hitting force causing impact on the PZT.

(B) Between period t1 andt2

The continue vibration of the PZT causing delay voltage oscillation with resonance frequency. However, this depends on type, structure and piezoelectric materials.

6.1 PZT System

The simulation and the experimental results of the PZT system will then be compared where the PSPICE was used for the simulation.

All the related parameters are calculated and its values are shown in Table 6.1. The value of inductor *Lre* and capacitor *Cre* are determined by the resonant frequency of the piezoelectric PZT and the data from (8). The piezoelectric PZT system is connected with five different resistive loads

Parameters	Values
Resistive Load (kΩ)	20, 50, 100, 200, 300
Applied force frequency (Hz)	0.5, 2
Resistance for coupling energy: $Rin1$	5 Ω
Resistance for coupling energy: $Rin2$	500 kΩ
Resistance for resonance: Rre	6 Ω
Internal resistance of the PZT: Rp	2 MΩ
Inductance for resonance: Lre	1305 μH
Capacitance for coupling energy: Cin	0.2 μF
Capacitance for resonance: Cre	3450 μF
Internal capacitance of the PZT: Cp	48.2 nF
Resonance frequency: f_0	75 Hz
Damping factor: ξ	0.059

Table 6.1: Parameter values [20,21]

6.2 Damping Factors

The voltage source was used as the proposed model. The main internal parameters are as following:

- the output resistance RP
- the output capacitance CP

They are measured by a RCL meter (Fluke PM6306).

The duration of the voltage response from Figure 6.2 can be modelled as a second-order parallel RLC circuit. The resonance equation can be express as a function of complex frequency (S1 and S2), which was used to determine the values of R, L and C.

Furthermore, the voltage response in this duration is found to be underdamped. Theoretically, the underdamped voltage response occurs when the damping factor, is less than one. The damping factor and the roots S1 and S2 can be expressed as:

$$\xi = \frac{\alpha^2}{\omega_0^2} = \frac{1}{2R}\sqrt{\frac{L}{C}}$$

$$s_{1,2} = -\alpha \pm j\omega_d$$

From analysis and the experimental results, the proposed model of studied piezoelectric

PZT system, the mechanical energy loss is modelled as series resistor *Rin1*.

6.3 System Modelling

Figure 6.3 shows the electromechanical model of the piezoelectric PZT. The energy from the input terminal to the output terminal of piezoelectric PZT is formed by the resistor *Rin2* and the capacitor *Cin*. The duration of the impact caused by the hitting force and the oscillation period can be controlled by switch SW1 and SW2.

These two switches are complementary. The switch SW1 closes when the piezoelectric PZT arm was hit and opened. Then, the piezoelectric PZT is releasing the energy while the switch SW2 is closed to emulate the resonated energy. A pulse voltage source, *Vpulse* is the representative of applied mechanical force. This frequency depends on how often the piezoelectric PZTwas hit by the mechanical force generator.

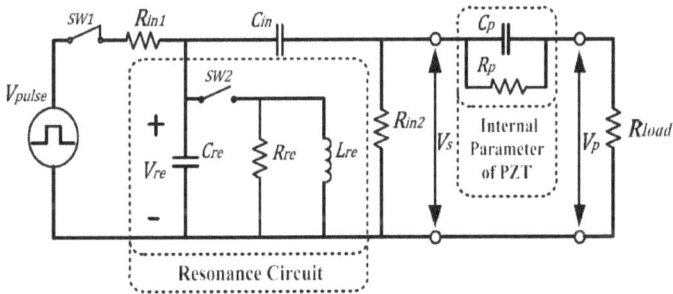

Figure 6.3: Electromechanical model of the piezoelectric PZT [20,21]

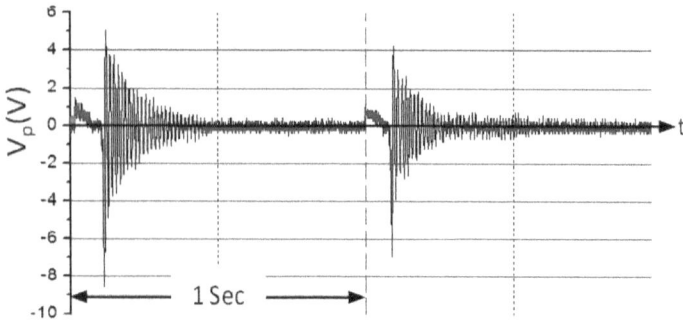

Figure 6.4: Voltage waveform of the PZT experiment results [20,21]

The experimental results shown in Figure 6.4 which was used to investigate by using Fast Fourier Transform (FFT). This is to identify its frequency components.

The behaviour of the piezoelectric PZT from the simulation and the experimental results are significantly similar, except no noise components in the simulation.

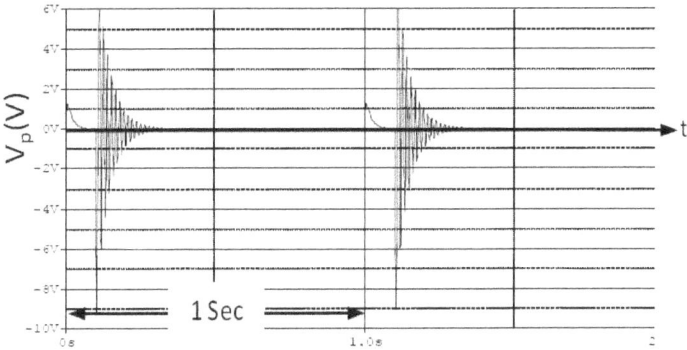

Figure 6.5: Voltage waveform of the simulation results from PSPICE [20,21]

6.4 Frequency Simulation

Table 6.2 shows the percentage error of the generated electrical energy. The proposed model was used to compare with the piezoelectric PZT prototype. In this particular comparison, the maximum error was 15.68% and minimum error was 0.26%.

R(kΩ) F(Hz)	20	50	100	200	300
0.5	10.02	15.68	-4.28	-2.03	-0.26
2.0	-13.52	-5.24	-12.21	-7.23	-7.04

Table 6.2: Percentage error of electrical energy [20,21]

Figure 6.6 shows the frequency spectrum and piezoelectric PZT voltage which data from the experiment results. Figure 6.7 shows the frequency spectrum and piezoelectric PZT voltage which data from the simulation results. These data were extracted from FFT in PSPICE program.

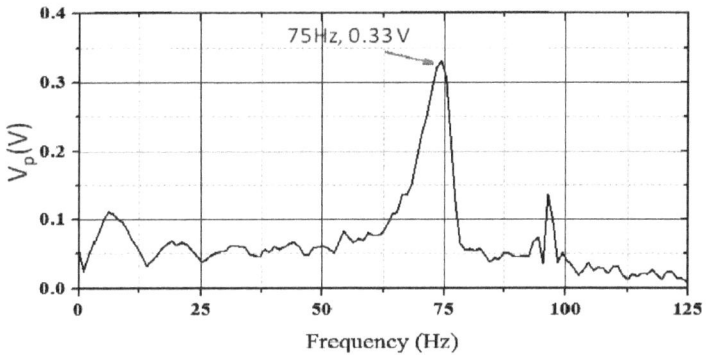

Figure 6.6. Frequency spectrum of PZT voltage(experiment results) [20,21]

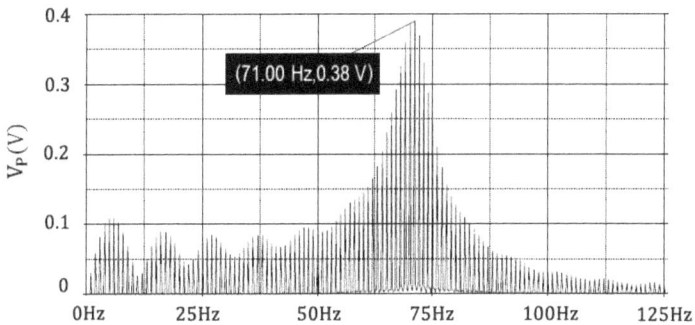

Figure 6.7: Frequency spectrum of PZT voltage (simulation results) [20,21]

The resonant frequencies of the experiment was at 75 Hz and the simulation was at 71 Hz. The percentage of error was approximately 5.34%.

6.5 Buck Converter

The model of piezoelectric PZT system to simulate the electrical energy harvesting system was shown in Figure 6.8 where the feedback controlled buck converter was used as the output voltage regulator.

The voltage generated by the piezoelectric PZT was rectified by a full-wave diode circuit. Then, later can be used in energy harvesting applications. After that, the unregulated DC voltage was feed back as the input of buck converter.

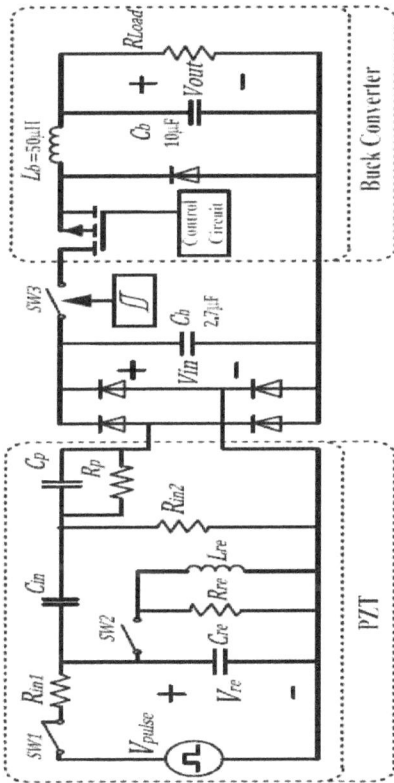

Figure 6.8: The simulated energy harvesting system with the proposed model [20,21]

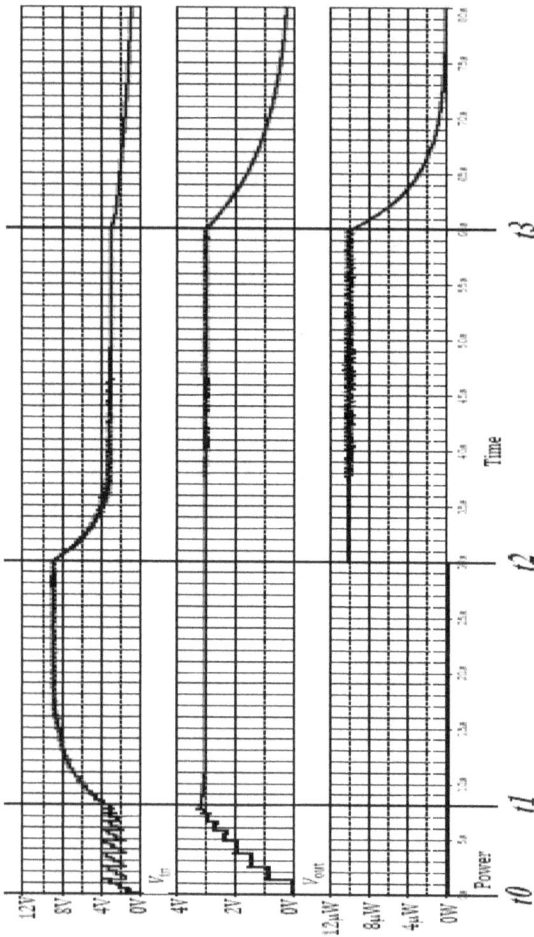

Figure 6.9: The simulation results of input voltage Vin, output voltage, Vout and resistive load power [20,21]

Piezoelectrics in Prosthetics: Energy Harvesting

Figure 6.9 shows the simulation results of input voltage Vin, output voltage, Vout and resistive load power. At *t0*, the mechanical force applied to the piezoelectric PZT with no electrical load. The electrical energy generated by the piezoelectric PZT which would start the accumulating in the capacitor, *Ch*.

Next, the capacitor voltage reaching 3.3V at *t1*, and the switch SW3 is closed, and the buck converter was started to operate and regulate the output voltage at 3.0V. The operation of the SW3 is controlled based on hysteresis loop. If the input was 2.5Hz, the steady state input voltage *Vin* was designed to set at 9.0 V.

At *t2*, a 900 k Ohms resistive load was connected to the buck converter. The power delivered to resistive load is 10 microwatts as shown in Figure 6.9.

The applied mechanical force would be stopped at *t3*. The electrical energy stored in *Ch* started to dissipate to the load causing *Vin* deceased. When *Vin* is lower than 3.3V, the SW3 would be turned off and the buck converter would stop operating. Voltage *Vout* was not regulated and delays to zero. The simulation results indicated that the proposed model could be practically utilized in the design process of an energy harvesting system.

This chapter proposed a new electro-mechanical modelling technique for PZT, which was employed in energy harvesting systems. The model was derived from a prototype's mechanical-to-electrical energy conversion characteristics.

The accuracy of the proposed model has been experimentally verified. The comparison results have indicated that the proposed model was consistent with the electrical characteristic of the studied PZT. It also demonstrated how the proposed model performs with the buck converter. The simulation results confirmed that the proposed model can be practically utilized

in the energy harvesting system design
process.

CHAPTER 7

DYNAMIC ANALYSIS

In an energy harvesting in the prosthesis, the energy harvest and storage element deforms during running or walking. Thus energy can be stored and released similar to that obtained from muscles. When a person wears the foot prosthesis, the foot and stump constitute a dynamic system.

Piezoelectric devices are capable of converting mechanical energy to electrical energy and vice visa. Many researchers are paying attention to harvest ambient vibration energy by using piezoelectric devices for micro power applications.

The output voltage of the piezoelectric devices is ac voltage, therefore, a power converter is necessary for converting the ac voltage to a dc voltage for supplying the electrical energy to loads those are usually electronic circuits.

To analyze the device, it must find a dynamic response of the foot to the action of a force similar to the force exerted by the ground by locomotion. In the analysis, the point connected to the socket will be fixed. [17]

The major limitations of harvesting the energy from piezoelectric devices are not only providing low power, but also the difficulty to investigate their behaviour when applying to some applications such as prosthetic legs.

Moreover, the behaviour of the piezoelectric devices would change when the operating conditions such as humidity or when it has to use to connect directly to a resistive load, and using with full-bridge diode rectifier, etc. The piezoelectric devices would act and perform in different ways. [22]

This would cause the utilization of the piezoelectric devices less efficiency that means high cost per Watt. It is, therefore, would be useful for energy management circuit analysis and design.

This chapter describes the effect of piezoelectric PZT structure with the open-circuit output voltage waveforms. The influent of the mechanical, electrical, and electromechanical were also described.

7.1 Lead Zirconate Titanate

The piezoelectric PZT and the force action on to the body structure was shown in Figure 7.1. This Piezoelectric PZT comprises two piezo-ceramic plates and two arms formed a single metal substrate [1-3, 10]. The piezo-ceramic is a high performance commercial grade Lead Zirconate Titanate formed.

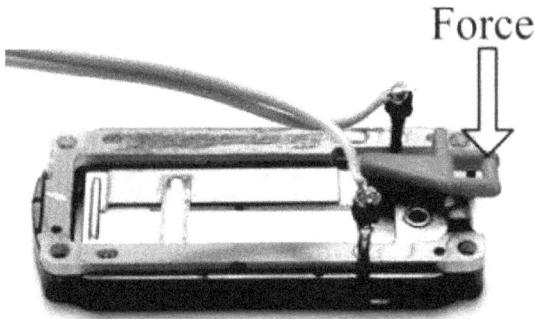

Figure 7.1:Piezoelectric PZT and force [1-3, 10, 19]

When applying mechanical force on the upper arm as shown in Figure 7.2 and ignoring energy loss. The energy is directly applied into the piezo-ceramic and converted to electrical energy while the rest is stored in the metal substrate in the form of potential energy.

7.2 Dynamic Structure

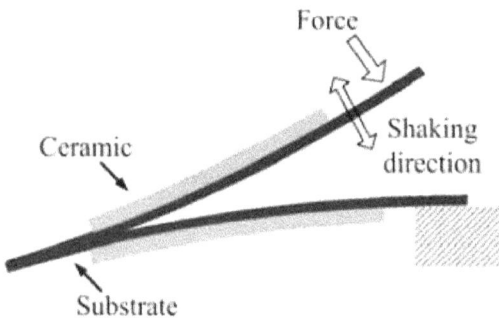

Figure 7.2: Dynamic structure of the PZT [1-3, 10, 19]

This results in the metal substrate and piezo-ceramic being continuously shake until the resonant period ends.

The electrical energy is continuously produced after stop applying the mechanical force in the resonant period.

7.3 Current Analysis

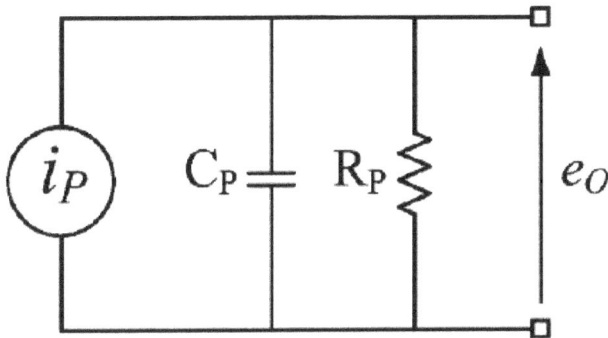

Figure 7.3. Piezoelectric PZT current [10, 19-21]

It shows the general output characteristic of the PZT. It can be either current source or voltage model, however, in this book the voltage source model is preferable.

7.4 Voltage Analysis

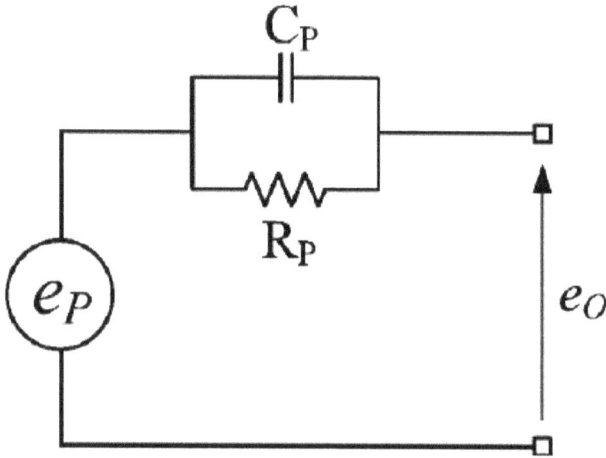

Figure 7.4: Piezoelectric PZT voltage {10, 19-21]

The experiment set up for investigating the output voltage waveforms when hitting the upper arm of the piezoelectric PZT.

7.5 Hit Dynamic

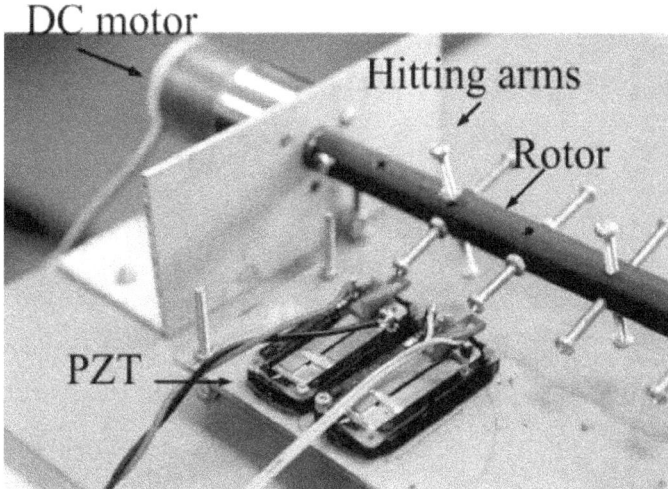

Figure 7.5: Experiment set up when hitting [1-3,19]

Figure 7.5 shows the experiment set up for investigating the output voltage waveforms when hit the upper arm of the piezoelectric PZT. The output voltage waveforms from this experiment setup was shown in Figure 7.7

7.6 Pull Dynamic

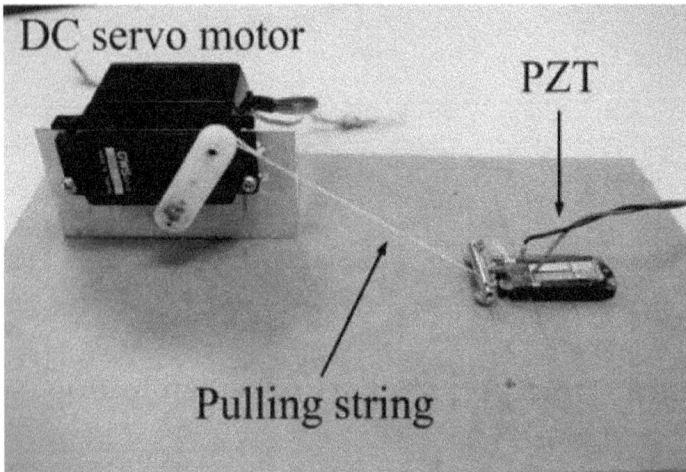

Figure 7.6: Experiment set up when pulling [1-3,19]

Figure 7.6 shows the experiment set up for investigating the output voltage waveforms when pull the upper arm of the piezoelectric PZT. The output voltage waveforms from this experiment setup was shown in Figure 7.8

7.7 Hit Experiments

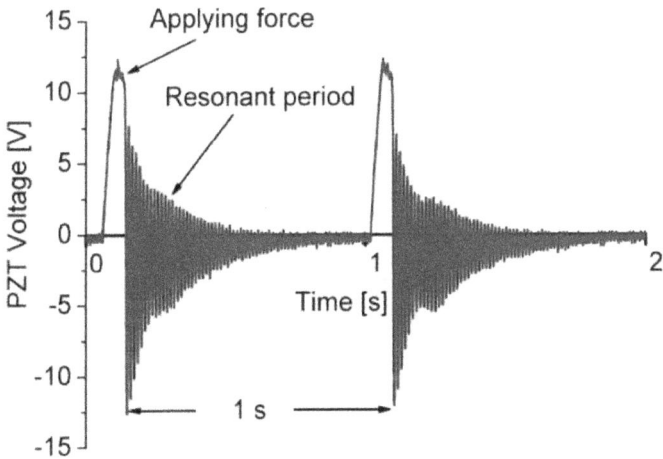

Figure 7.7: Output voltage waveforms hitting [19-21]

7.8 Pull Experiments

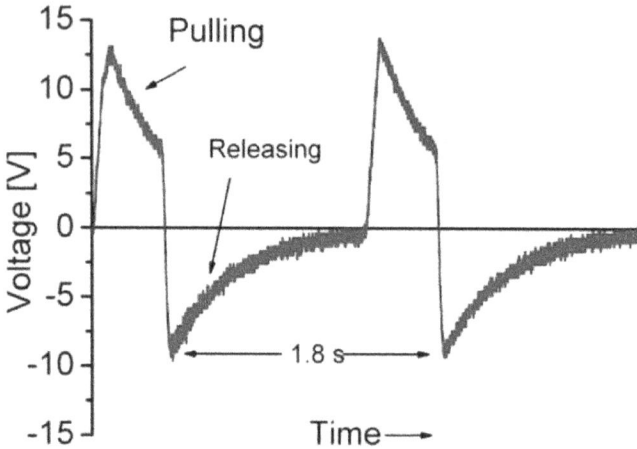

Figure 7.8: Output voltage waveforms pulling [19-21]

In an energy harvesting prosthesis the energy harvest and storage element deforms during running or walking.

Thus energy can be stored and released similar to that obtained from muscles.

When a person wears the foot prosthesis, the foot and stump constitute a dynamic system.

To analyze the device, it must find a dynamic response of the foot to the action of a force similar to the force exerted by the ground by locomotion. In the analysis, the point connected to the socket will be fixed. The forcing function depends on the individual and also on the speed of locomotion. When the time of the peak value of ground force changes from 0.09 to 0.15 sec., the propulsion force varies from 62% to 85%. [17]

Piezoelectrics in Prosthetics: Energy Harvesting

CHAPTER 8

SIMULATION AND EXPERIMENTS

This chapter presents a modelling method based on PSCAD/PSPICE, which emulates the behaviour of the piezoelectric PZT. The investigation of the actual open-circuit output voltage at no load operation together with measuring the output parameters (resistance and capacitance) of the PZT by RLC meter were applied and the data were used to complete the modelling by PSCAD/PSPICE.

This model was able to operate in variety conditions. Therefore, it would be useful not only for the energy management circuit analysis but also in the design system.

This simulation and experimental results of the piezoelectric PZT were studied. The output voltage waveforms was used to discussed with the general output characteristic of the piezoelectric PZT from the simplified mechanical model [23]. The mechanical and electrical models of the piezoelectric PZT can be written as shown in Figure 8.1 and Figure 8.2.

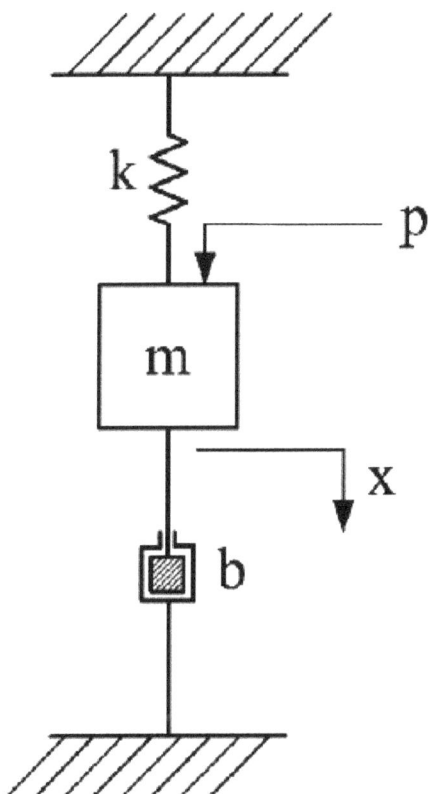

Figure 8.1: The Simplified mechanical model [1-3, 19]

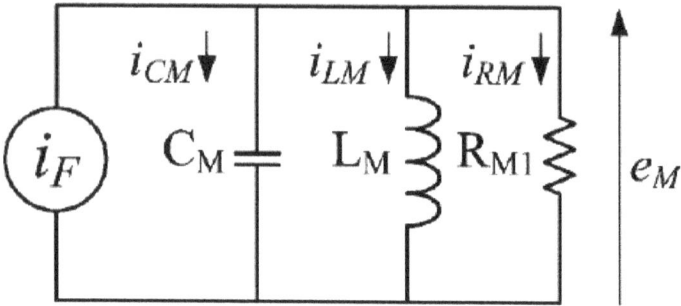

Figure 8.2: Mechanical and electrical

Model [1-3, 10]

The electromechanical model that includes

mechanical and electrical models can be

obtained. In Figure 8.2, the output

parameters: CP and RP can be measured by

a RLC meter.

The mechanical and electrical parameters:

CM, LM, and RM1 can be defined by

Piezoelectrics in Prosthetics: Energy Harvesting

investigating the open-circuit output voltage waveform at the resonant period and assumingly is a second order resonant voltage waveform.

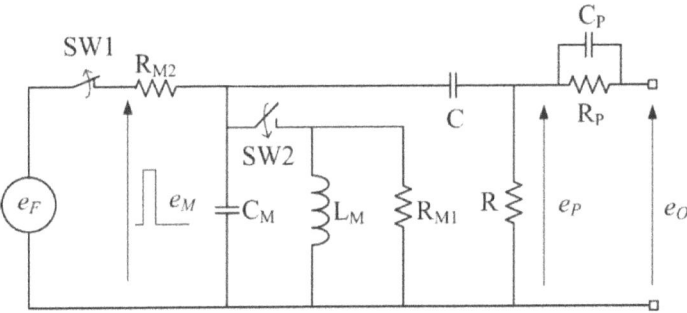

Figure 8.3: The electromechanical model of the PZT [1, 19]

Therefore, the damping factor of the resonant voltage waveform is defined by:

$$\xi = \frac{1}{2R_M}\sqrt{\frac{L_M}{C_M}}$$

For the hit states: the damping factor can be as low as 0.06. On the other hand, in the case of pulling, the damping factor is as high as 1.5. The resonant frequency is defined by:

$$f_0 = \frac{1}{2\pi\sqrt{L_M C_M}}$$

Piezoelectrics in Prosthetics: Energy Harvesting

The resonant frequency was approximately 80 Hz where the RM2 is the representative of mechanical energy loss and set at 5 Ohm. C and R are for coupling the energy form input side to the output side of the piezoelectric PZT. This could use to find out the ramping down time of the output voltage during the period of being hit or pulled.

The SW1 and SW2 are complementary. The SW1 would close the circuit when the PZT's arm was hit or pulled. The SW1 would open the circuit when stop hitting or releasing the PZT's arm.

The eF is the representative of mechanical force. The value could be approximately set a bit higher than the peak of the open-circuit output voltage. There are two parameters those must be changed when the type of mechanical force changes as follows: RM1 = 5 for hitting force and = 0.2 for pulling force.

The *eM* is a rectangular voltage waveform that has the pulse width of 0.1s when the period of 1s for hitting force. In addition, he pulse width of 0.5s when the period of 1.8s for pulling force. The resonant frequency is set by LM and CM from (10, 19-21).

8.1 Operating Waveforms

In this book, the simulation can be obtained by PSCAD or PSPICE simulation software.

8.2 Hit Simulation

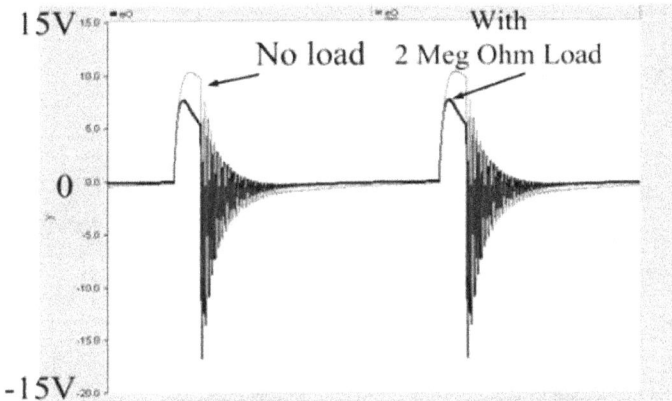

Figure 8.4: PZT voltage waveforms with hitting force: Simulation [19-21]

Figure 8.4 shows the simulation while Figure 8.5 shows the experiments results of the piezoelectric PZT voltage waveforms with hitting force at the frequency of hitting at 1 Hz (nearly the frequency of walking) when directly connecting a 2-MΩ resistive load and at no load.

8.3 Hit Experiments

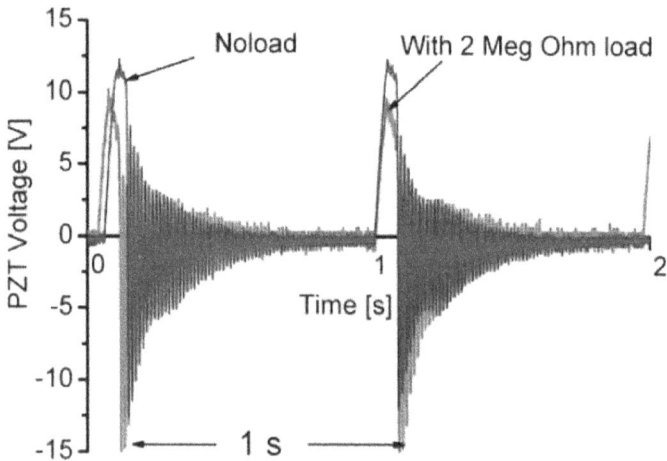

Figure 8.5: PZT voltage waveforms with hitting force: Experiments [19-21]

8.4 Pull Simulation

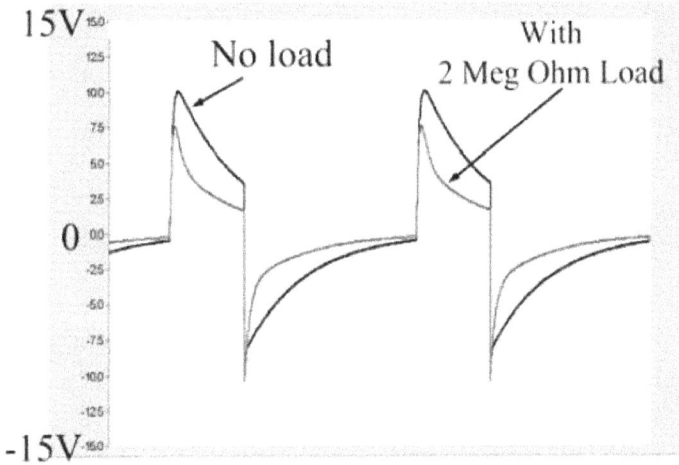

Figure 8.6: PZT voltage waveforms with pulling force: Simulation [19-21]

The open-circuit voltage waveform of the PZT with pulling force seems to have higher rms voltage. However, with a resistive load connection, the voltage is much reduced than the case of hitting force.

8.5 Pull Experiments

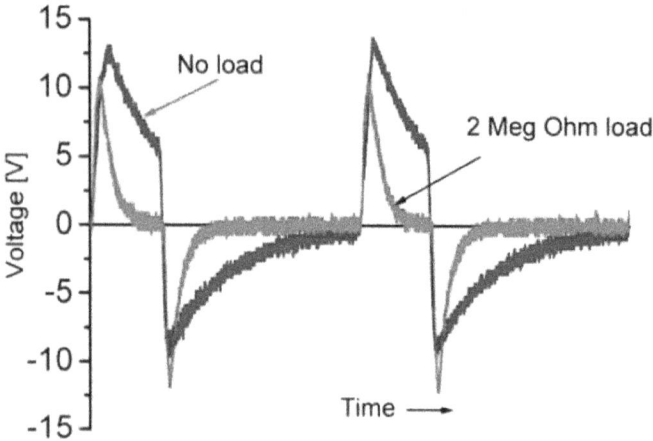

Figure 8.7: PZT voltage waveforms with pulling force: Experiments [19-21]

Figure 8.7 shows the results of applying pulling force at the frequency of 0.6 Hz. It is clear that the simulation results agree well with the experimental results from Figure 8.5

The piezoelectric PZT output power with the pulling force is much lower than the hitting force.

Piezoelectrics in Prosthetics: Energy Harvesting

CHAPTER 9

CONVERTER AND RECTIFIER

This chapter presents PZT (Lead Zirconate Titanate) intended for using in low frequency mechanical movement applications. The output voltage of the piezoelectric devices is ac voltage. Therefore, a power converter is necessary for converting the ac voltage to dc voltage for supplying the electrical energy to loads which usually are electronic circuits.

From Chapter 6-8, the simplified piezoelectric PZT and electromechanical modelling based on PSCAD/PSPICE, the simulation and experiment can emulate the behaviour of the PZT in variety conditions. The simulation and experimental results well agree with each other. The model is able to operate in variety conditions. It is, therefore, would be useful for energy management circuit analysis and design.

The disadvantage of the energy harvesting from piezoelectric devices are providing low power and the operating conditions. Thus, the operating cost is quite high compare in term per Watt energy production.

This book describes the modelling of the PZT by investigating the output voltage of the PZT and using simplified mechanical, mechanical-to-electrical, and electromechanical models.

9.1 Buck and Boost

The power converter buck and boost are widely used in widely applications [24] for converting a dc voltage level to another level of the dc voltage and including for regulation voltage required. Their output is dc voltage that is suitable for general electronics control circuits.

When the input power source is an ac voltage and connect to the full-bridge diode rectifier for converting an ac voltage to a dc voltage. This can been done as a solution for using with the buck and boost converters.

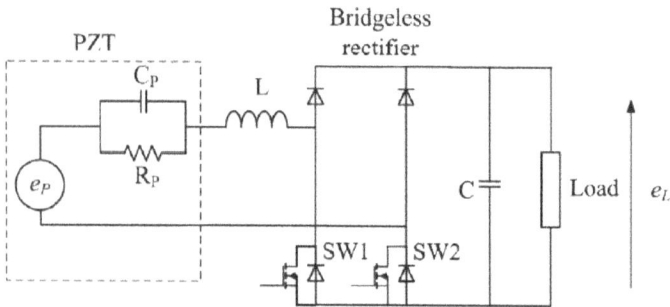

Figure 9.1: Bridgeless rectifier for energy harvesting circuit [1-3, 19]

Figure 9.1 shows the details circuit of the bridgeless rectifier for energy harvesting circuit. It consists of an inductor, two switching devices, two diodes, and one output capacitor. This rectifier [25] can rectify an ac voltage from the input to a dc voltage at the output. It also can regulate the output voltage.

However, in the case of the piezoelectric PZT, using the full-bride diode rectifier is not an effective way for harvesting the energy. The buck and boost converters are not suitable for using as the energy harvesting circuits for the piezoelectric PZT.

The operating system started with a switching device turns on. Thus, the energy from the piezoelectric PZT can be directly stored in the inductor. Next it delivered to the output via a diode and then the switching device would be turns off.

During the system was in the status of turn on position, the voltage would drop low the circuit system. When using the MOSFET as the switching devices even the energy is delivered to the output via a diode during the turn off period, the voltage drop at the diode would not affect the maximum power at the piezoelectric PZT. This is because the diode was installed in the output circuit section. Therefore, the bridgeless rectifier is more suitable for energy harvesting circuits for the piezoelectric PZT. This would exact higher power from the PZT.

9.2 Load Connection

The experiments and simulation to study the piezoelectric PZT show that both the simulation and experimental results agree well with each other.

Figure 9.2 shows the power and the Rms output voltage curves of the piezoelectric PZT when the resistive load is directly connect.

The Power and Vrms for both simulation and experimental data travelled along in the route until the power reached at 25 μW and the Voltage Vrms reached at approximately 5 Volts, the simulation and the experiments started to travel separately.

The simulation power and the simulation voltage Vrms met at the load resistance at 0.62 Mega Ohm when the simulation power was approximately 15 µW and the simulation voltage was about 3 Volts.

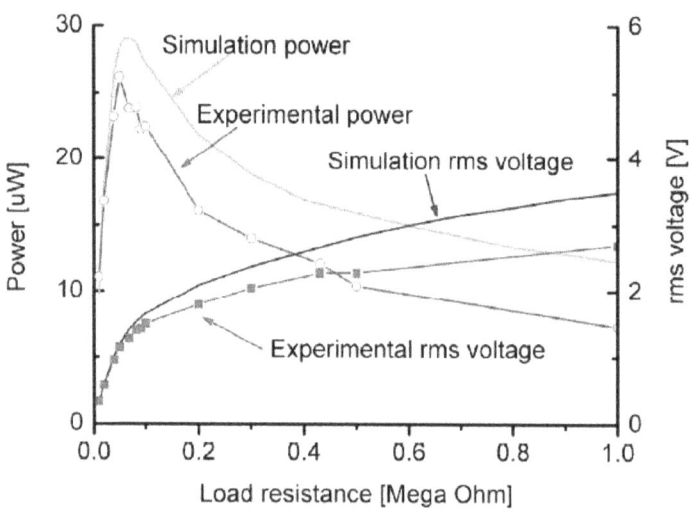

Figure 9.2: Rms voltage and power curves of the PZT with hitting force (1Hz) when directly connected to the resistive load [19]

However, the experimental results the power and the Vrms was met at the load resistance at 0.48 Mega Ohm. The power at that point was approximately 10 μW and the voltage was about 2.5 Volts.

At the simulation power shows that the maximum power point is approximately 29 μW while the experimental power is approximately 27 μW. The maximum power of the simulation is higher than the experimental results.

9.3 Full-Bridge Diode Rectifier

The energy harvesting circuit with full-bridge diode rectifier was shown in Figure 9.3. This is a simple circuit which consists of fours diodes and one capacitor. During the resonant period, the resonant voltage has an instantaneous voltage level lower than the average output voltage.

This was because the diode rectifier cannot deliver the energy to the output. It can consider that the full-bridge diode rectifier was not the best and effective way to harvest the energy from piezoelectric PZT.

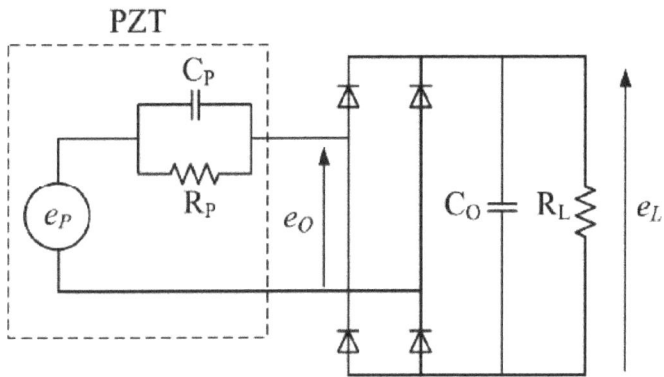

Figure 9.3: Full-bridge diode rectifier energy harvesting [19-21]

9.4 PZT Full-bridge Rectifier

Figure 9.4 shows the DC output voltage and output power curves of the piezoelectric PZT when it was hit at frequency 1 Hz and it connected to the full-bridge diode rectifier. These simulation and experimental results were obtained from

low frequency mechanical force at 1 Hz
which this is a human walking frequency.

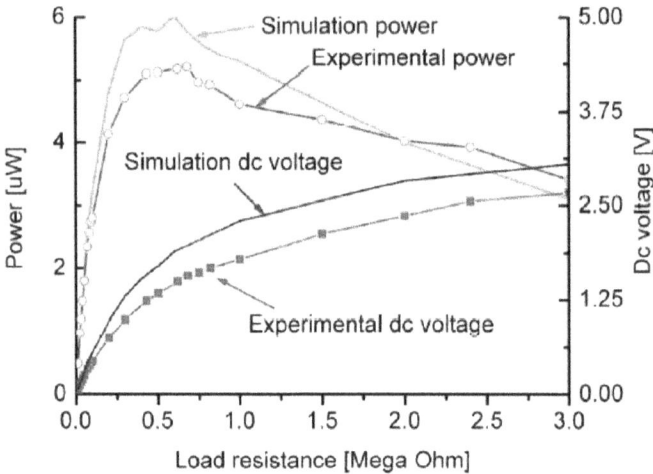

Figure 9.4: DC output voltage and output
power curves of the PZT with hitting force
(1 Hz) when connected with the full-bridge
diode rectifier [19]

The maximum power from simulation was
much lower with full-bridge diode rectifier

than without. The maximum power was about 6 μW with the full-bridge diode rectifier while the maximum power was 29 μW when connected directly to resistive load.

The results from the experiments shows the same profiles to the simulation. The maximum power from the experimental results was approximately 5 μW.

The results both from simulation and experiments can confirm it is possible to emulate the behaviour of the PZT in various conditions. However, the maximum power generated from the piezoelectric PZT could be designed to make it suitable for every kinds of energy.

This includes to the energy management system prior to achieve the maximum power.

CHAPTER 10

MFC

In this chapter, the characteristic of the piezoelectric material, Macro Fibre Composites (MFC), has been investigated. Nowadays, there are high growth of electronics industry. Most of electronics parts and circuits have been applied widely in every places. Therefore, alternative energy supply sources were an inevitable component in those systems or applications.

The comparison between the electrical equivalent circuit based simulation and the experimental results were observed.

The operational factors, internal impedance and frequency could affect the maximum power output of the piezoelectric. The rectifier circuit for the energy harvesting system and the most suitable energy storage were also considered. Both experiments and measurements were used on our laboratory energy harvest kit compare data to the commercial MFC.

10.1 Background

In this book, a conventional Electro-Chemical battery has been considered to use as an energy storage. However, there are few various disadvantages such as a limited of capacity, lifecycle, weight, size and leakages of the chemical solution. In addition, the continuously maintenance and reliability for the security monitoring.

In order to solve these problems, the energy harvesting and energy scavenging were widely interested by the researchers around the world.

Among various types of energy harvester equipments, a piezoelectric material is one of the most effective component which convert the surrounding mechanical vibration energy into the electrical energy directly.

As it has been known for many decades, the piezoelectric materials have been used as sensors, switching devices in circuit breakers and actuators. [1-3, 36-48]

10.2 Problems

Although the mechanical vibrations take place in most of the structures. However, the energy harvested performance of the piezoelectric material was not very big. Moreover, the level of the energy generated was very small in the unit of micro-watt or milli-watt order [1-3, 10, 29].

However, this could be done with some improvement on the piezoelectric material and fabrication process. This includes the low energy consumption from the electronics circuit design. Therefore, it is possible to use the piezoelectric to harvest more energy in various kinds and ranges of their applications such as medical implanted devices [30], diagnosis equipment for large power plant [31], power source for remote wireless sensing nodes [32-33], ultra low power ubiquitous applications [34] and in low frequency application such as prosthetic hand [35].

10.3 Macro Fibre Composite

A Macro Fibre Composite (MFC) is an alternative piezoelectric material which offers high performance and flexibility in a cost competitiveness [1-3, 27-28]. It has very flexible and durable structure. To consider the energy harvesting performance, the mechanical part and electrical part have to be investigated. Also, the characteristic of each part including the energy harvesting systems have to be analysed.

10.4 MFC Modelling

The piezoelectric energy harvesting performance could be evaluated on different Electro-Mechanical platform and computer modelling [19-21, 26]. Those models considered the effect between mechanical part and electrical part simultaneously. There was no data or research about the electrical characteristic of the MFC piezoelectric clearly mentioned before.

10.5 Equivalent Circuit

The Macro Fibre Composite (MFC) can be represented as an ideal energy source (voltage or current source) with internal resistance as shown in Figure 10.1 [10, 19-21]

When considering at the linear source and load, the Maximum Power Point (MPP) would be occurred only at the operating point which the external load resistor (RL) equals to the internal resistor (Ri) in the piezoelectric material. However, it was also reported that this MMP values could be changed due to operating frequency.

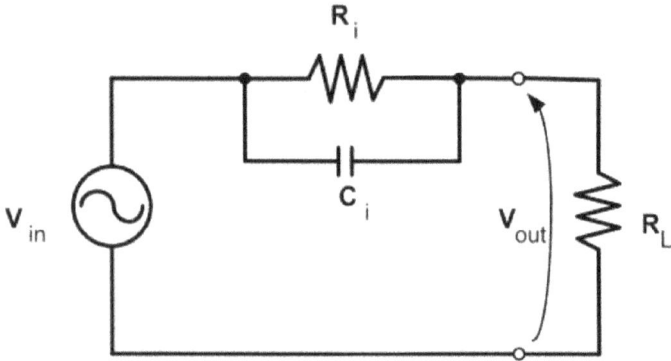

Figure 10.1: Equivalent circuit of MFC [19-21, 26]

Figure 10.1 shown the electrical equivalent circuit of the MFC where Ri and Ci are the open circuit value of internal resistant and internal capacitance of the MFC.

The values of resistance and capacitance could be measured by the RLC meter at the operating frequency region.

The energy source AC sinusoidal voltage connected directly to the Ri and Ci. [19-21, 26]. The maximum power would be occurred when the external impedance equals to internal impedance. When the load resistor RL was connected to the equivalent circuit of MFC, the maximum power would be considered. Therefore, the equivalent circuit was the total internal impedance, ZIn_Total can be expressed as the relation between impedance from resistance, $ZR= Ri$ and impedance from

capacitance, $Z_C = 1/\omega C_i$ as following
equations:

$$Z_{In\_Total} = \frac{Z_C Z_R}{Z_C + Z_R}$$

$$= \frac{R_i}{1 + R_i C_i * 2\pi f}$$

From the equation, the maximum power
would be changed corresponding to the
operating frequency. The higher the
operating frequency, the larger value of
impedance of capacitance..

At operating frequency region, the value of
the load resistance that cause the maximum

power point *RL_peak* can be approximated as following equation: [29]

$$R_{L\_peak} = Z_{In\_Total} = \frac{1}{C_i * 2\pi f}$$

The value would be change inversely to the operating frequency and the value of capacitance.

In this Chapter the electrical parameter, the output voltage of the MFC and the evaluating method on electrical model were studied. The commercial energy harvesting kit was used as a testing platform. Both the experimental and simulation results focused on the maximum power of the MFC at each testing condition are investigated. The effect from the internal resistance in selected capacitor within the energy harvesting circuit was also continue investigated in Chapter 11 and Chapter 12.

CHAPTER 11

ENERGY STORAGE

In this chapter the characteristic of the piezoelectric material Macro Fibre Composites (MFC) has been investigated. They are the electrical parameter. The energy harvester piezoceramic can convert the mechanical energy directly into the electrical energy.

The energy harvesting circuit usually composes of mechanical and electrical part. At each testing condition the effect an internal resistance and capacitance in the energy harvesting circuit were invested in this book.

11.1 Energy Storage

The power generated from the energy harvesting was used to store in the energy storage tanks which could be capacitor banks. It was also used to supply directly to the load. However, this depends on how the application required.

The conventional circuit which always used for energy harvesting circuit as shown in Figure 11.1. It started with the MFC would generate ac power. Then, this power was rectified by the full bridge rectifier circuit. After the certain period of time the energy would be accumulated into the capacitor where the energy would be stored.

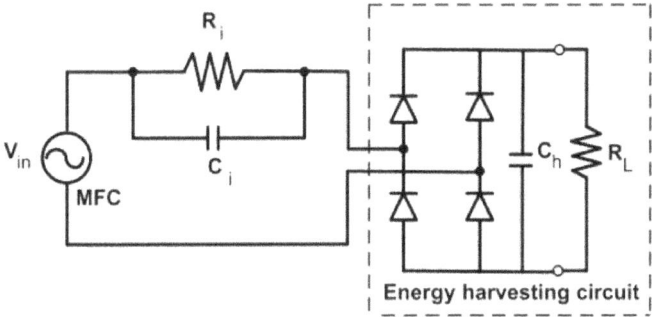

Figure 11.1: Energy harvesting circuit

The Energy can be define as following:

$$w(t) = \frac{1}{2}cv^2(t)$$

The quantity of the energy was depends on the capacitor size. This shows the capability energy from the energy harvesting. Nevertheless, the voltage level could also affect the stored energy. Therefore, the capacity energy from the energy harvesting could be upon the output voltage as well as the capacitor size.

11.2 MFC Output

This book investigated the value of the capacitor and types of the capacitors such as Tantalum, Electrolyte capacitor and Niobium oxide. Each capacitor also has different internal impedance which could be also directly affect to the leakage current

or the capacity to maintain the output
voltage.

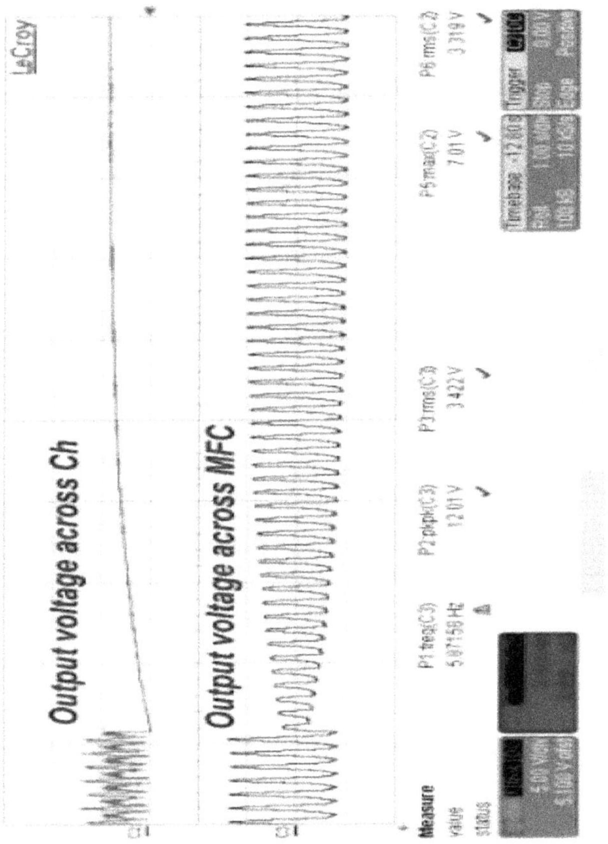

Figure 11.2: MFC output voltage [1,14, 19-
21,26]

Figure 11.2 shows the voltage waveforms across the MFC and *Ch*. When the energy harvesting circuit is connected to the MFC, the voltage across *Ch* will increase continuously to some certain level. The response of the output voltage across *Ch* will be faster when the frequency of the voltage waveform of the MFC is high. Also, with larger value of internal resistance in *Ch*, the voltage across *Ch* will also increase faster.

11.3 Capacitor Output

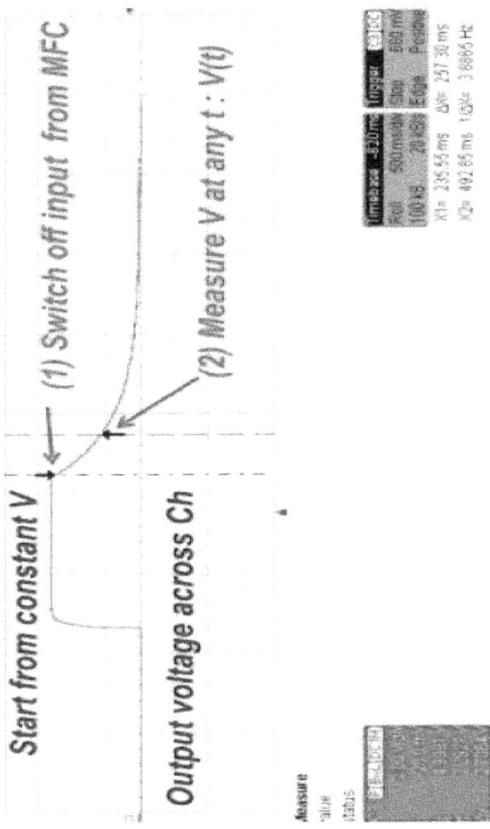

Figure 11.3: Capacitor output voltage [1,14-21,26]

Therefore in some application liked energy harvesting in prosthetic leg which the frequency only 1-2 Hz, the larger value of internal resistance in Ch is preferable, since the output voltage could be accumulated to the required level before it would be leaked out or be discharged to the load. With different type, value and voltage rating of each capacitor, the internal impedance will be different. Consequently, the leakage current will also be different.

The internal resistance of each capacitor can be determined from the simple test. The capacitor and the load resistance which is connected in parallel to Ch will be charged to any certain voltage level by energy circuit or DC supply. Then at any time, the input is switched off in order to let the stored energy in *Ch* discharge to the load resistance as shown in Fig. 4.

The decreasing ratio of the output voltage or the time constant that output voltage decreases from same certain voltage level, can be defined by its capacitance value and internal resistance as following equation.

$$R = \frac{-t}{C * \ln(\frac{v(t)}{V})}$$

11.4 Energy Harvesting Test

The experimental setup where the commercial smart energy harvester development kit from Smart material Corp is used. It composes of a shaker which its frequency can be adjusted from 0 to 60 Hz and control system for adjusting the testing conditions.

In this book, a P2 type d31 coupling mode as shown in Fig. 5 is used as a specimen. Although the coupling efficiency of this type is lower than the other modes, it is one commonly used since it is most efficient for the small force and low vibration level environment. In this type, a force is applied in the direction perpendicular to the poling direction. The tested MFC model MFC8528 is mounted on the flexible arm and connect to the energy harvesting kit.

Figure 11.4: MFC structure and details [1, 10, 19-21,26]

The end of the flexible arm with MFC on the top is clamped to the periodical moving pole in the central point of the harvesting kit while on another end of the flexible arm, a 18 grams of weight is attached in order to keep the constant movement of the flexible arm and constant deforming of the MFC or

a constant mechanical energy input

condition.

Model	MFC 8528
Type	P2 (d31 mode)
Size	8.5*2.8 cm
Maximum operational positive voltage	+ 360 V
Maximum operational negative voltage	- 500 V
Operational bandwidth	< 10 kHz

Table 11.1: MFC Specification

Figure 11.5: MFC and vibration kit [1, 19-21,26]

CHAPTER 12

POWER HARVESTING

A piezoelectric material is one effective component to convert the surrounding mechanical vibration energy into the electrical energy directly. From prior literature during these decades the piezoelectric materials have long been used as sensors and actuators.

Although the mechanical vibrations take place in most of the structures, but the possible harvested performance of the piezoelectric material is not high and the level of the energy is very low in micro-watt or milli-watt order [1].

However with some improvement on the piezoelectric material and fabrication process as well as the low energy consumption of the electronics circuit design makes the possibility for using the piezoelectric in more wide range applications.

A Macro Fibre Composite (MFC) is an alternative piezoelectric material which offers high performance and flexibility in a cost competitiveness.

In order to optimize the performance of whole energy harvesting circuit which usually composes of mechanical part and electrical part, the characteristic of each part including the whole system have to be analytically determined.

Those models considered the complex decoupling effect between mechanical part and electrical part simultaneously. In this book, with the fixed electrical parameter, i.e. output voltage of the tested MFC, an evaluating method based on only electrical equivalent model is focused in this article. The commercial energy harvesting kit is used as a testing platform.

Both the experimental and simulation results focused on the maximum power point of the MFC at each testing condition are investigated. Furthermore, the effect from the internal resistance in selected capacitor within the energy harvesting circuit is also considered in this book.

12.1 Frequency and Voltage

The maximum power point at each operating frequency is determined by varying the load resistance.

The amplitude of movement of the shaking pole of the energy harvesting kit is adjusted in order to keep the output voltage from the MFC to be the same 10 Vpeak-peak.

12.2 Voltage Waveform

It indicates the waveform of the open circuit output voltage across the MFC. With the symmetrical bending of the flexible arm, the MFC is also symmetrically deformed up and down.

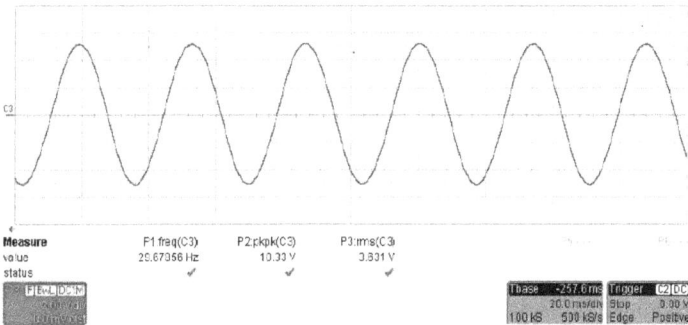

Figure 12.1: Measured waveform of output voltage of MFC [1, 19-21,26]

It is seen that the open circuit voltage of a MFC is a pure sinusoidal function of time, so the voltage source in Chapter 10 and Chapter 11 could be represented by a usual sinusoidal function too.

12.3 Load Resistance

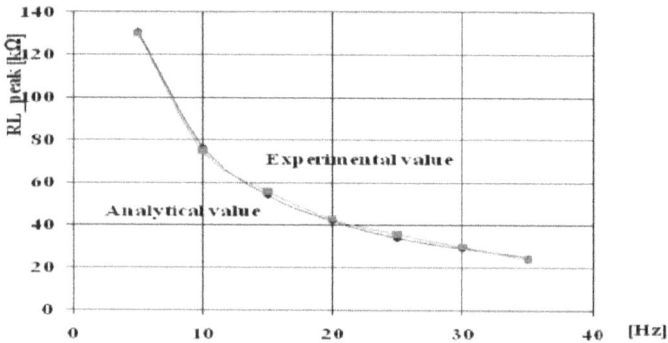

Figure 12.2: Relation between *RL_peak* and frequency [1, 19-21,26]

In order to verify the effectiveness of the proposed equivalent circuit in Chapter 10 and Chapter 11, the simulation with the

Piezoelectrics in Prosthetics: Energy Harvesting

same testing condition such as open circuit voltage from MFC and frequency etc., has been investigated by PSCAD program and the results are compared to the experimental r results.

The relation between the value of load resistance caused a maximum power point (*RL-peak*) and frequency. It is found that the simulation results agree well to the experimental results.

Furthermore, in the higher frequency region, with the effect from the impedance of capacitance, the value of *RL-peak* becomes smaller than the value in lower frequency. With the operating frequency limitation of the energy harvesting kit about 60 Hz, the results in the higher operating frequency cannot be determined, therefore the equivalent circuit model will be useful for checking the operation in the other applications which have higher operating frequency.

12.4 Power Harvesting

It shows the relation among load resistance, power and operating frequency. The harvested power is calculated from the measured voltage value by digital oscilloscope and measured value of load resistance directly. It is seen that the higher the operating frequency, the higher the power that could be harvested from the MFC.

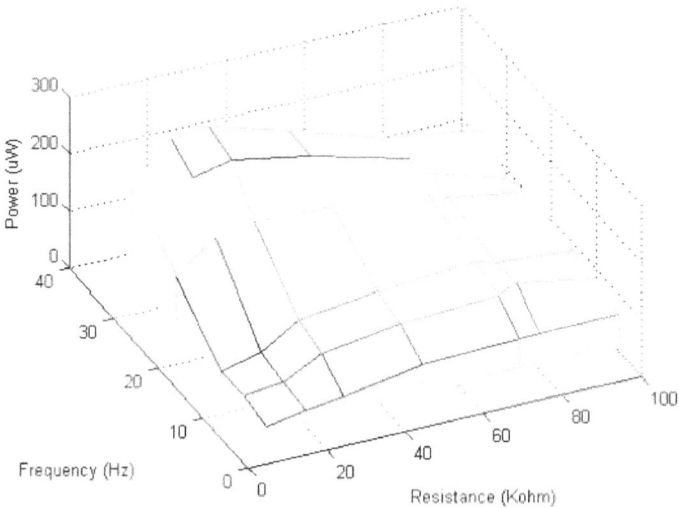

Figure 12.3: Relationship between resistance, frequency and power [1, 19-21,26]

12.5 Internal Resistance

Table 12.1 shows the measured value of internal leakage resistance of the capacitor inside the energy harvesting circuit (*Ch*) as shown in chapter 10.

The Niobium oxide trends to have a larger value of internal resistance. Similarly, the internal resistance of each type of capacitor which have different capacitance value are also measured. [19-21, 26]

Type	Neo Oxide	Electrolytic	Polypropylene	Niobium Oxide
Capacitor / Voltage Rating	4.7μF / 63V	4.7μF / 16V	4.7μF / 250V	4.7μF / 10V
Resistance (MΩ)	10.89	10.00	10.65	11.28

Table 12.1: Measured value of internal resistance in each type of capacitor

Electrolytic	Capacitor / Voltage Rating	47μF / 100V	4.7μF / 100V	0.47μF / 100V	
	Resistance (MΩ)	8.82	9.34	10.01	
Niobium Oxide	Capacitor / Voltage Rating	47μF / 10V	10μF / 10V	4.7μF / 10V	2.2μF / 10V
	Resistance (MΩ)	9.91	11.27	11.28	12.09
Polyster Film (WIMA)	Capacitor / Voltage Rating	10μF / 63V	10μF / 63V	10μF / 63V	10μF / 63V
	Resistance (MΩ)	9.58	9.84	10.55	10.86

Table 12.2: Measured value of internal resistance in different capacitance

The internal resistance becomes larger when the value of capacitance is smaller. However, the internal resistance of the electrolytic capacitor that has different rating voltage is also measured. Its value will be also varied by the rating voltage.

Electrolytic	Capacitor / Voltage Rating	4.7µF / 16V	4.7µF / 25V	4.7µF / 35V	4.7µF / 100V
	Rresistance (MΩ)	10.00	10.97	10.14	9.34

Table 12.3: Measured value of internal resistance for different rating voltage

With the effect from impedance of internal capacitor in the MFC, the value of the load resistance that causes the maximum power becomes smaller when the operating frequency becomes higher too.

The effect from the capacitor in the harvester circuit is also considered. With the suitable selection of capacitor which has high internal impedance, the energy will be stored effectively with low leakage.

CHAPTER 13

CONCLUSION

This book present the design of an energy harvesting system using smart materials for self power generation of upper and lower prosthetic legs. The green energy harvesting here will emphasis use of piezoelectric devices in Prosthetic legs. Geometric parameters, beam, mass and resistive electric loads significantly influence the output power.

The smart materials like Piezo Composites, Piezo Flexible Film, Macro Fibre Composites, and PZT have been employed and modified to be appropriately embedded in the prosthesis.

The movements of the prosthesis would extract and transfer energy directly from the piezoelectric via a converter to a power management system. Afterward, the power management system manages and accumulates the generated electrical energy to be sufficient for later powering electronic components of the prosthesis.

A new electro-mechanical modelling technique for PZT, which was employed in energy harvesting systems. The model was derived from a prototype's mechanical-to-electrical energy conversion characteristics.

The accuracy of the proposed model has been experimentally verified. The comparison results have indicated that the proposed model was consistent with the electrical characteristic of the studied PZT. It also demonstrated how the proposed model performs with the buck converter. The simulation results confirmed that the proposed model can be practically utilized in the energy harvesting system design. The experimental results of energy harvesting and efficiency in peak piezoelectric voltages during step up and continuous walking for a period of time and it is one of the most interesting sources for energy harvesting from environmental vibrations.

The devices that have been used are piezoelectric, electromagnetic, electrostatic, pyroelectric, photovoltaic and thermoelectric. The conversion of harvesting energy is very good in scalability, capability, high energy density and compatible with standard electronic technology. In addition, the piezoelectric could be coupled to a mechanism to perform opening of the contacts in the switching devices.

The piezoelectric CMOS harvesting could bypass the input voltage and recover some energy to increase the energy during negative piezoelectric voltage. With the effect from impedance of internal capacitor in the MFC, the value of the load resistance that causes the maximum power becomes smaller when the operating frequency becomes higher too.

The effect from the capacitor in the harvester circuit is also considered. With the suitable selection of capacitor which has high internal impedance, the energy will be stored effectively with low leakage.

The basic characteristics of the MFC piezoelectric has been investigated in this book by comparison between electrical equivalent circuit based simulation and experimental results. Experiments have been carried out on commercial energy harvesting kit. The condition of maximum power point at each operating frequency was focused.

References

[1] Book, "Piezoelectrics in Circuit Breakers: Design & Test", by Dr Kesorn Pechrach Weaver, ISBN-13: 978-0993117800, ISBN-10: 0993117805, Pechrach Publishing, 2004.

[2] K. Pechrach, J.W. McBride, P.M. Weaver, Arc root mobility on piezo actuated contacts in miniature circuit breakers, 49th IEEE Holm Conference on Electrical Contacts, Washington DC, USA, September 2003.

[3] P.M.Weaver, K.Pechrach, J.W.McBride, "Arc Root Mobility on Piezoelectrically Actuated Contacts in Miniature Circuit Breakers", IEEE Transactions on Components, Packaging, and Manufacturing Technologies, Vol. 28, No.4, December 2005, pp. 734-740.

[4] M. Zhu, E.Worthington, "Design and testing of piezoelectric energy harvesting devices for generation of higher electric power for wireless sensor networks", IEEE Sensors 2009.

[5] D. Kwon, G.A. Rincon-Mora, " A rectifierfree Piezoelectric Energy Harvester Circuit", IEEE 2009.

[6] P.M. Weaver, A. Ashwell, Y. Zheng, S.C. Powel, "Extended temperature rang piezo actuator system with very large movement", SPIE 1oth Annual International Symposium on Smart Structure and Materiald, San Diego, March 2003.

[7] Ceramtec handbook, " Smart material advanced Piezo Composites".

[8] J.M.Donelan, Q. Li, et al, "Biomecahnical energy harvesting", vol 319, Science.

[9] J. Andrysek, "Generator for prosthesis and orthosis", United States Patent No. 259320

[10] K. Pechrach, P. Manooonpong, F. Woegoetter, K. Tungpimolrut, N. Hatti, J.Phontip, K. Komol, "Piezoelectric Energy Harvesting for Self Power Generation of Upper and Lower Prosthetic Legs", Piezo2011, Sestriere, Italy, Feb - March 2011.

[11] S. C. Powell, "The development of a low cost, high performance piezo actuator for high volume manufacture," in Proc. 7th Int. Conf. New Actuators (Actuator'00), Bremen, Germany, Jun. 2000, pp. 45–48.

[12] P. M. Weaver, S.C. Powell, "LOW COST PRECISION CONTROL OF PIEZOCERAMIC ACTUATORS", Actuator 2002, Proc. 8th International Conference on New Actuators, Bremen, 10-12 June 2002

[13] K. Pechrach, "Arc Control in Circuit Breakers with low Contact Velocity", TCP2010 Green Thailand, Bangkok,

Thailand, July 2010.

[14] K. Pechrach, "Electroceramics Green Energy Harvesting in Industrial Estates and Agriculture in Thailand", 12th ATPER 2009 - The Association of Thai Professionals in Europe, Paris, France, May 2009.

[15] Loreto Mateu and Francesc Moll, "System-level simulation of a self-powered sensor with piezoelectric energy Harvesting", 2007 International Conference on Sensor Technologies and Applications, page 399-404.

[16] A. Kansal, D. Potter, and M. Srivastava. Performance aware tasking for environmentally powered sensor networks. In Joint International Conference on Measurement and Modeling of Computer Systems, 2004.

[17] [100] Walter D. Pilkey, William A. Gruver, Dewen Jin, Jichuan Zhang, "Analytical Evaluation of an Energy-Storing Foot Prosthesis",IEEE, 1995.

[18] J. R. Liang and W. H. Liao, "Energy flow in piezoelectric energy harvesting systems," J. Smart Mater. Struct., vol.20, no.1, act. no. 105005 (11 pp.), January 2011.

[19] N. Hatti, K. Tungpimolrut , J.Phontip, K. Pechrach, K. Komol, P. Manooonpong, " PZT Energy Harvesting and Management Circuits for Prosthetic Legs ", ISAF-PFM-2011, Vancouver, Canada, July 2011.

[20] K. Komol, S. Sirisukpraser, K. Tungpimolrut, N. Hatti, K. Pechrach, P. Manooonpong, , "Modeling and Analysis of a PZT Piezoelectric for Energy Harvesting System", ECTI-CON 2012, Hua Hin, Thailand, May 2012.

[21] K. Komol, S. Sirisukpraser, K. Tungpimolrut, N. Hatti, K. Pechrach, P. Manooonpong, , " Electrical Modeling of Piezoelectric Element for Energy Harvesting Circuits", EECON-34, Chonburi, Thailand, November 2011.

[22] G. K. Ottaman, A. C. Bhatt, H. Hofmann, and G. A. Lesieutre, "Adaptive piezoelectric energy harvesting circuit for wireless remote power supply," IEEE Trans. Power Electron., vol. 17, pp. 669--676, Sept. 2002.

[23] K. Ogata, "Modern Control Enineering," Prentice Hall International Editions, United States of America, 1990.

[24] N. Mohan, T. M. Undeland, and W. P. Robbins, "Power Electronics: Converters, Applications, and Design" Wiley International Edition, United States of America, 1989.

[25] L. Huber, J. Yungtaek, and M. M. Jovanovic, "Performance evaluation of bridgeless PFC boost rectifier," IEE

[26] K. Tungpimolrut, N. Hatti, J.Phontip, K. Komol, K. Pechrach, P. Manooonpong, " Design of Energy Harvester Circuit for a MFC Piezoceramic based on Electrical Circuit Model Optimization ", ISAF-PFM-2011, Vancouver, Canada, July 2011.

[27] K. Pechrach, "Piezoelectric Force Microscopy (PFM) Materials Characteristics for Medical Application", ATPER 2011, Copenhagen, Denmark, May 2011.

[28] K. Pechrach, "Electroceramics Green Energy Harvesting in Industrial Estates and Agriculture in Thailand", 12th ATPER 2009 - The Association of Thai Professionals in Europe, Paris, France, May 2009.

[29] Hyun-Uk Kim, Woo-Ho Lee, H. V. Rasika Dias, and Shashank Priya, "Piezoelectric Microgenerators - Current Status and Challenges," IEEE Tran. on Ultrasonics, Ferroelectrics, and Frequency Control, vol. 56, no. 8, pp. 1555-1568, 2009.

[30] Stephen R., Platt, Shane F., Kevin G. and Hani H., "The Use of Piezoelectric Ceramics for Electric Power Generation Within Orthopedic Implants," IEEE/ASME Trans. on Mechatronics, vol. 10., no.4, pp. 455 – 461, 2005.

[31] Bartosz P. A., Piotr P., Maciej M. and Andrzej N., "Enhancement of Piezoelectric Vibration Energy Harvester Output Power Level for Powering of Wireless Sensor Node in Large Rotary Machine Diagnostic System ," IEEE International Conf. on Mixed Design of Integrated Circuits and Systems, June 2009.

[32] Geffrey K. Ottman, Heath F. Hofmann, Archin C. Bhatt, and George A. Lesieutre, "Adaptive Piezoelectric Energy Harvesting Circuit for Wireless Remote Power Supply," IEEE Trans. on Power Elect., vol. 17, no. 5, pp. 669 – 676, 2008.

[33] Y.K. Tan, J.Y. Lee and S.K. Panda, "Maximize Piezoelectric Energy Harvesting Using Synchronous Charge Extraction Technique For Powering Autonomous Wireless Transmitter," IEEE ICTES, pp. 1123 – 1128, 2008.

[34] Lu Chao, Chi-Ying Tsui and Wing-Hung Ki, "A Batteryless Vibrationbased Energy Harvesting System for Ultra Low Power Ubiquitous Applications," IEEE International Symposium on Circuits and Systems, pp. 1349 -1352, 2007.

[35] Darryl P. J. C, Paul H. C., Andy N. M W. and Steve P. B., "A Novel Thnick-Film Piezoelectric

Slip Sensors for a Prosthetic Hand," IEEE Trans. on Sensors Journal, vol. 7, No. 5., May 2007

[36] Book, "Slow Contact Opening Circuit Breakers", by Dr Kesorn Pechrach, ISBN-13: 978-0993117862, ISBN-10: 0993117864, Pechrach Publishing, 2015.

[37] Book, "Arc Control in Circuit Breakers: Low Contact Velocity", by Dr Kesorn Pechrach, ISBN-13: 978-3639221015, VDM Publishing house, 2009.

[38] P.M.Weaver, K.Pechrach, J.W.McBride, "The Energetics of Gas Flow and Contact Erosion During Short Circuit Arcing", IEEE Transactions on Components and Packaging Technologies, Vol. 27, No.1, March 2004, pp. 51-56.

[39] J.W.McBride, K.Pechrach, P.M.Weaver, "Arc Motion Gas Flow in Current Limiting Circuit Breakers Operating with a Low Contact Switching

Velocity", IEEE Transactions on Components, Packaging, and Manufacturing Technologies, Vol. 25, No.3, September 2002, pp. 427-433.

[40] J.W.McBride, K.Pechrach, P.M.Weaver, "Arc Root Commutation from Moving Contacts in Low Voltage Devices", IEEE Transactions on Components and Packaging Technologies, Vol. 24, No.3, September 2001, pp. 331-336.

[41] K. Pechrach, "Arc Control in Circuit Breakers with low Contact Velocity", TCP2010 Green Thailand, Bangkok, Thailand, July 2010.

[42] K. Pechrach, "Electroceramics Green Energy Harvesting in Industrial Estates and Agriculture in Thailand", 12th ATPER 2009 - The Association of Thai Professionals in Europe, Paris, France, May 2009.

[43] K. Pechrach, "Modeling of Renal Vascular System for Nanomedical Applications", Franco-

British Nanomedicine Summit Countdown to FP7, Paris, France, November 2006.

[44] K. Pechrach, J.W. McBride, P.M. Weaver, Arc root mobility on piezo actuated contacts in miniature circuit breakers, 49th IEEE Holm Conference on Electrical Contacts, Washington DC, USA, September 2003.

[45] K.Pechrach, J.W.McBride, P.M.Weaver, "The Correlation of Magnetic, Gas dynamic and Thermal Effects on Arc Mobility in Low Contact Velocity Circuit Breakers", 48th IEEE Holm Conference on Electrical Contacts, Florida, USA, October 2002, pp.86-94.

[46] K.Pechrach, J.W.McBride, P.M.Weaver, "Analysis of Arc root Mobility in Low Contact Opening Velocity Circuit Breakers", 21st ICEC/ITK 2002 International Conference on Electrical

Contacts, Zurich, Switzerland, September 2002, pp.260-267.

[47] K.Pechrach, J.W.McBride, P.M.Weaver, "Gas Flow and Composition Effects on Arc Motion in Current Limiting Circuit Breakers", 47th IEEE Holm Conference on Electrical Contacts, Montreal, Canada, September 2001, pp.12-17.

[48] J.W.McBride, K.Pechrach, P.M.Weaver, "Arc Root Commutation from Moving Contacts in Low Voltage Devices", 46th IEEE Holm Conference on Electrical Contacts, Piscataway, NJ, USA, September 2000, pp.130-138.

[49] Jacopo Carpaneto, Silvestro Micera, Franco Zaccone, Fabrizio Vecchi, and Paolo Dario, "A Sensorized Thumb for Force Closed-Loop Control of Hand Neuroprostheses", IEEE TRANSACTIONS ON NEURAL SYSTEMS AND

REHABILITATION ENGINEERING , VOL. 11,
NO. 4, DECEMBER 200, page 346-353.

[50] M. M. Adamczyk and P. E. Crago, "Input-output nonlinearities and time delays increase tracking errors in hand grasp neuroprostheses," IEEE Trans. Rehab. Eng., vol. 4, pp. 271–279, Dec. 1996.

[51] M. C. F. Castro and A. Cliquet, Jr, "A low-cost instrumented glove for monitoring forces during object manipulation," IEEE Trans. Rehab. Eng., vol. 5, pp. 140–147, June 1997.

Index

ac voltage, 126, 162, 166

accuracy, 82, 83, 122, 234

across, 205, 221

action, 47, 92, 98, 127, 130, 141

activate, 48

adhesive, 9

advancing, 33

analysis, 87, 90, 106, 127, 129, 141, 145, 164

analytical, 78

angle, 7, 19, 48, 55, 69

ankle, 37, 52, 53, 78

application, 31, 187, 200, 207

applications, 84, 90, 116, 126, 128, 162, 165, 182, 187, 216, 225

applying, 128, 131, 133, 159

arms, 6, 17, 88, 130

assembled, 23

attach, 48

attractive, 66

ball of the foot, 58, 59, 61

beam, 4, 16, 19, 21, 86, 232

beginning, 60

behavior, 82, 83, 110, 127, 128, 144, 163, 180

behaviors, 82, 83, 85, 88, 90

bend, 48, 49, 51

bimorph, 8, 16, 18, 19, 20, 21

boost, 165, 166, 168

bridge diode rectifier, 175, 179

Bridgeless, 167

buck, 14, 83, 115, 116, 120, 121, 123, 165, 166, 168, 235

Buck Converter, 115

cables, 47

cantilever, 16, 86

cantileverbeam, 86

capability, 4, 202, 236

capacitance, 104, 144, 192, 193, 194, 195, 199, 210, 224, 227, 229

capacitor, 28, 45, 63, 71, 102, 107, 119, 120, 167, 174, 196, 199, 200, 201, 202, 208, 209, 219, 227, 228, 229, 230, 231, 238

capacitor banks, 45, 63, 199

ceramic, 7, 9, 11, 17, 18, 22, 24, 81, 88, 130, 132, 133

ceramic lead zirconate titanate, 81, 88

ceramics, 25

changes, 53, 142, 152

characteristic, 91, 97, 123, 134, 145, 182, 189, 190, 198, 217, 235

characteristics, 85, 88, 91, 122, 234, 239

circuit, 12, 13, 43, 45, 62, 81, 87, 95, 100, 104, 115, 129, 144, 145, 148, 152, 157, 163, 167, 169, 174, 182, 183, 186, 187, 191, 193, 196, 198, 200, 201, 205, 209, 216, 217, 219, 220, 221, 223, 225, 227, 231, 238, 239

coefficient, 22

coercive, 26

commercial, 22, 131, 184, 196, 210, 218, 239

compact, 17

compare, 112, 164, 183

comparision, 112

complementary, 108, 151

complex, 25, 36, 104, 218

component, 95, 182, 185, 215

conditions, 30, 128, 144, 163, 164, 180, 211

connections, 23

consistent, 123, 235

constitutive, 29

consumption, 31, 43, 187, 216

contacts, 4, 98, 237

contract, 53

controlled, 107, 115, 120

conversion, 3, 89, 122, 234, 236

converter, 2, 14, 83, 115, 116, 120, 121, 123, 126, 162, 165, 233, 235

converting, 12, 125, 126, 162, 165, 166

coupling, 151, 211

current, 67, 71, 72, 99, 134, 190, 203, 208

damping factor, 105, 149, 150

decreases, 27, 210

deflection, 16, 27, 29

deformed, 221

deforming, 213

delay, 44, 101

density, 4, 237

design, 1, 21, 82, 122, 123, 129, 145, 164, 187, 216, 231, 235

development, 15, 84, 210

device, 31, 85, 127, 141, 168

diagram, 12, 13, 18, 29

different, 18, 33, 60, 78, 103, 128, 189, 203, 208, 227, 229, 230 diode, 115, 128, 166, 168, 169, 174, 175, 176, 177, 178, 179

direction, 28, 36, 53, 69, 212

directly, 2, 9, 14, 128, 132, 155, 168, 171, 173, 179, 185, 192, 198, 199, 203, 215, 225, 233

displacement, 21, 22, 30, 93

duration, 104, 105, 107

dynamic, 87, 94, 125, 127, 141

effect, 25, 53, 78, 129, 189, 196, 199, 218, 219, 224, 230, 231, 237, 238

efficiency, 3, 14, 89, 129, 211, 235

elastic, 52

electric, 5, 30, 32, 84, 232

electrical, 2, 12, 15, 23, 30, 66, 82, 83, 85, 87, 90, 91, 93, 94, 95, 99, 111, 112, 115, 119, 121, 122, 123, 125, 126, 129, 132, 133, 146, 147, 148, 162, 165, 182, 185, 188, 189, 191, 195, 198, 215, 217, 218, 233, 234, 235, 239

electrode, 9, 22

electromagnetic, 3, 236

electromechanical, 97,
107, 130, 147, 149,
163, 165

electro-mechanical,
81, 97

electro-mechanical,
122

electro-mechanical,
234

Electromechanical,
109

electronics, 166, 182,
187, 216

electrostatic,, 3, 236

element, 28, 52, 125,
140

embed, 42

employed, 2, 81, 85,
97, 122, 233, 234

employs, 16

emulate, 82, 90, 94,
108, 163, 180

emulates, 144

energy, 1, 2, 3, 4, 5, 12,
14, 15, 17, 18, 19, 20,
21, 22, 24, 26, 27, 28,
31, 43, 45, 47, 52, 55,
57, 61, 64, 66, 68, 71,
73, 75, 76, 78, 81, 82,
84, 85, 86, 89, 92, 97,
99, 106, 107, 108,
112, 115, 116, 117,
119, 121, 122, 123,
125, 126, 127, 129,
132, 133, 140, 141,
145, 151, 162, 163,
164, 167, 168, 169,
174, 175, 176, 180,
182, 183, 184, 185,
186, 187, 188, 189,
190, 192, 196, 198,

199, 200, 201, 205,
207, 209, 210, 212,
213, 215, 216, 217,
218, 219, 220, 224,
227, 231, 232, 233,
234, 235, 236, 237,
238, 239

energy tanks, 45

epoxy, 11

equivalent circuit, 193

error, 111, 112, 114

evaluated, 189

experiment, 110, 113,
114, 135, 137, 138,
163

experimental, 3, 68,
71, 90, 102, 106, 110,
145, 159, 163, 170,
171, 173, 177, 179,

183, 196, 210, 218,
223, 224, 235, 239

experiments, 97, 154,
170, 171, 179, 180,
183

expressed, 93, 105,
193

fabrication, 187, 216

feed back, 116

film, 9, 10, 28, 56, 60,
62, 64

flexible arm, 212, 213,
221

foot, 12, 13, 34, 52, 55,
56, 57, 58, 59, 60, 61,
62, 63, 64, 65, 68, 69,
70, 71, 73, 74, 75, 76,
77, 78, 125, 127, 141

foot area, 65

foot prosthesis, 78,
  125

force, 16, 21, 27, 37,
  53, 65, 98, 99, 101,
  107, 108, 119, 121,
  127, 130, 131, 133,
  141, 152, 153, 154,
  155, 156, 157, 158,
  159, 172, 177, 178,
  211

forces sensors, 65

formed, 7, 17, 22, 107,
  130

frequency, 88, 89, 98,
  101, 102, 104, 108,
  110, 113, 150, 151,
  153, 155, 159, 162,
  177, 183, 187, 191,
  192, 194, 195, 205,
  208, 211, 219, 222,
  223, 224, 225, 226,
  230, 238, 239

full wave, 14

Full-Bridge, 174

function, 52, 104, 142,
  222

generated, 2, 9, 51, 67,
  82, 98, 99, 100, 111,
  115, 119, 180, 186,
  199, 233

generator, 108

Geometric, 4, 232

geometry, 16

geothermal, 84

grain, 22

ground, 36, 37, 52, 53,
  56, 59, 74, 76, 78,
  127, 141

growth, 182

harvesting, 1, 3, 4, 5, 12, 13, 15, 17, 18, 19, 20, 21, 22, 24, 26, 27, 31, 36, 37, 39, 40, 41, 42, 43, 45, 46, 47, 49, 51, 57, 58, 62, 66, 67, 68, 70, 71, 73, 75, 76, 81, 82, 83, 85, 115, 116, 117, 122, 123, 125, 127, 140, 164, 167, 168, 170, 174, 183, 184, 188, 189, 196, 198, 199, 200, 201, 202, 205, 208, 212, 213, 217, 218, 219, 220, 224, 227, 231, 232, 234, 235, 236, 237, 239

high cost, 129

hit or pulled, 151, 152

hitting, 42, 43, 99, 101, 107, 136, 139, 152,

153, 154, 155, 156, 158, 159

horizontal, 69

hysteresis, 25, 120

impact, 101, 107

impedance, 86, 183, 192, 194, 203, 208, 224, 230, 231, 237, 238

implants, 66

improvement, 186, 216

index, 53

individual, 142

inductor, 102, 167, 168

industrial, 84

influent, 129

input, 5, 107, 116, 118, 119, 120, 151, 166, 167, 209, 213, 237

installing, 71

interdigitated, 9, 10

interfaced, 34

internal, 103, 183, 190, 191, 192, 196, 199, 203, 205, 208, 209, 210, 219, 227, 228, 229, 230, 231, 237, 238

interval, 31

inversely, 195

investigate, 110, 127

investigated, 15, 86, 98, 182, 188, 196, 198, 202, 219, 223, 239

investigating, 135, 137, 138, 148, 164

joint, 37, 53, 78

knee-joint, 34

laboratory, 183

Lead Zirconate Titanate, 6, 97, 130, 131, 162

leakage, 203, 208, 227, 231, 238

level, 165, 175, 186, 202, 205, 208, 209, 210, 211, 216

limitation, 224

limitations, 127

load resistance, 209, 225

loads, 5, 126, 162, 232

locomotion, 127, 141

loss, 14, 81, 106, 132, 151

lower, 1, 27, 70, 72, 76, 77, 121, 159, 175, 179, 211, 224, 232

magnets, 15

management, 2, 129, 145, 163, 180, 233

material, 5, 28, 42, 52, 182, 185, 186, 187, 188, 191, 198, 211, 215, 216, 217

materials, 25, 101, 185, 215

maximum, 26, 44, 65, 67, 74, 76, 87, 112, 169, 173, 178, 179, 180, 183, 192, 194, 195, 196, 219, 223, 230, 238, 239

means, 18, 31, 129

measured, 98, 104, 148, 192, 225, 227, 228, 229

mechanical, 28, 30, 81, 85, 87, 92, 93, 94, 95, 98, 106, 108, 119, 121, 122, 125, 129, 131, 133, 145, 147, 148, 151, 152, 162, 165, 177, 185, 186, 188, 189, 198, 213, 215, 217, 218, 234

mechanical force, 108, 152

mechanical strain, 30

mechanical stress, 30

mechanism, 4, 15, 16, 24, 27, 43, 45, 48, 237

metal, 7, 17, 23, 130, 132, 133

metalized, 28

MFC, 9, 10, 13, 14, 67, 70, 71, 72, 73, 74, 181, 182, 184, 188, 189, 190, 191, 193, 195, 198, 200, 202, 204, 205, 212, 213, 214, 217, 218, 219, 220, 221, 223, 226, 230, 238, 239

microcontroller, 26, 31

model, 38, 81, 82, 83, 87, 103, 106, 107, 109, 112, 115, 117, 121, 122, 134, 144, 145, 147, 149, 163,

196, 212, 218, 225, 234

modulus, 52

motion, 16, 37, 53

moulded, 42

movement, 7, 12, 13, 16, 19, 162, 213, 220

moving, 69, 213

muscles, 53, 79, 125, 141

natural, 66, 93

negative, 5, 52, 54, 237

neutral, 31

opening, 4, 237

operating, 17, 121, 128, 164, 168, 190, 192, 194, 195, 219,

224, 225, 230, 238, 239

operating frequency, 194, 224, 226

oscillation, 101, 107

output, 5, 21, 75, 78, 86, 98, 104, 107, 115, 118, 119, 120, 126, 129, 134, 135, 137, 138, 144, 145, 148, 151, 152, 159, 162, 164, 165, 167, 169, 171, 175, 176, 178, 183, 195, 202, 203, 204, 205, 207, 208, 209, 218, 220, 221, 232

output voltage, 137, 139, 167, 210

parameter, 91, 195, 198, 218

parameters, 4, 25, 83, 96, 102, 104, 144, 148, 152, 232

peak, 3, 67, 71, 72, 74, 75, 76, 77, 142, 152, 195, 220, 222, 223, 224, 235

percentage, 111, 114

perfect, 34

performance, 25, 26, 86, 130, 186, 188, 189, 215, 217

performs, 82, 123, 235

period, 3, 36, 52, 53, 101, 107, 133, 148, 151, 153, 169, 174, 200, 236

periodical, 213

perpendicular, 28, 212

phenomenon, 29

piezo, 7, 9, 30, 130, 132, 133

piezoceramic, 10, 25, 198

piezoelectric, 2, 3, 4, 5, 6, 8, 9, 14, 15, 17, 22, 24, 25, 28, 29, 45, 47, 51, 56, 57, 60, 62, 63, 64, 67, 68, 71, 73, 74, 81, 85, 88, 89, 91, 93, 94, 97, 101, 102, 106, 107, 108, 109, 110, 112, 113, 115, 119, 126, 127, 128, 129, 130, 136, 137, 138, 144, 145, 151, 154, 159, 162, 163, 164, 168, 169, 170, 171, 175, 176, 180, 182, 183, 185, 186, 187, 188, 189, 191, 198, 215, 216, 217, 232, 233, 236, 237, 239

Piezoelectric, 5, 6, 8, 24, 27, 41, 43, 47, 49, 61, 70, 74, 86, 125, 130, 131, 134, 135

planar, 8, 15, 18, 20, 21

plates, 7, 17, 18, 23, 88, 130

platform, 68, 189, 196, 218

point, 8, 20, 127, 141, 173, 190, 195, 213, 219, 223, 239

position, 25, 26, 27, 38, 40, 48, 49, 50, 51, 55, 56, 57, 65, 71, 75, 78, 169

positions, 55, 60, 61, 77

positive, 54

potential energy, 132

power, 2, 5, 9, 31, 38, 45, 52, 53, 57, 63, 66, 67, 70, 72, 74, 75, 76, 77, 78, 84, 86, 87, 119, 120, 126, 127, 159, 162, 164, 165, 166, 169, 171, 172, 173, 176, 178, 179, 180, 183, 187, 192, 194, 195, 196, 199, 200, 219, 223, 225, 226, 230, 232, 233, 238, 239

power management, 2, 233

processing, 22

production, 92, 164

promising, 86

proposed, 81, 82, 83, 94, 103, 106, 112, 117, 121, 122, 223, 234

propulsion, 52, 142

prosthesis, 2, 12, 13, 34, 52, 68, 69, 70, 71, 73, 77, 78, 125, 140, 141, 233

prosthetic, 1, 12, 13, 24, 37, 39, 41, 42, 47, 48, 49, 51, 55, 57, 60, 62, 63, 64, 65, 68, 70, 128, 188, 208, 232

Prosthetic knee, 46

prosthetic legs, 1, 12, 13, 68, 70, 128, 232

prototype, 82, 83, 86, 100, 112, 122, 234

PSPICE, 83, 102, 111, 113, 144, 153, 163

Pull Dynamic, 137

PVDF, 28

pyroelectric, 3, 236

PZT, 1, 5, 6, 8, 12, 14, 44, 45, 47, 67, 68, 71, 72, 81, 82, 86, 87, 88, 89, 90, 91, 94, 97, 99, 100, 101, 102, 106, 107, 108, 109, 110, 112, 113, 114, 115, 119, 122, 123, 129, 130, 131, 133, 134, 135, 136, 137, 138, 144, 145, 149, 151, 152, 154, 155, 156, 157, 158, 159, 162, 163, 164, 168, 169, 170, 171, 172, 175, 176, 178, 180, 233, 234, 235

ramping, 151

rating, 208, 229, 230

ratio, 209

reaction, 37, 51, 53

rectangular, 9, 16, 153

rectifier, 14, 128, 166, 167, 168, 170, 174, 175, 176, 177, 178, 179, 183, 200

redesign, 15

regulation, 165

relative, 37, 53

relative motion, 54

released, 52, 125, 141

releasing, 108, 152

represent, 93

researchers, 126, 185

residuum, 34

resistance, 87, 104,
144, 172, 173, 190,
192, 193, 195, 196,
199, 205, 208, 209,
210, 219, 220, 223,
225, 226, 227, 228,
229, 230, 238

resistive load, 119,
120, 128, 155, 157,
171, 173, 179

resistive loads, 103

resonance, 94, 101,
104

resonant, 81, 102, 114,
133, 148, 149, 150,
151, 153, 174

resonant voltage
waveform, 148

response, 87, 94, 99,
100, 104, 105, 127,
141, 205

result, 87, 100

results, 3, 21, 71, 77,
82, 83, 89, 90, 102,
106, 110, 111, 113,
114, 118, 119, 121,
123, 133, 145, 154,
159, 163, 170, 173,
174, 177, 179, 180,
183, 196, 218, 223,
224, 235, 239

revealed, 87, 100

reverse field, 26

rms, 157

roots, 105

running, 84, 125, 141

scalability, 4, 236

self power, 1, 232

sensors, 65, 185, 215

series resistor, 106

setup, 68, 99, 137, 139, 210

shaker, 211

shaking, 220

shoe, 28

Simplified, 147

simulated, 83, 117

simulation, 51, 102, 110, 111, 113, 114, 118, 119, 121, 123, 145, 153, 154, 159, 163, 170, 171, 173, 177, 178, 179, 180, 182, 196, 218, 223, 224, 235, 239

single, 7, 17, 19, 21, 130

Smart, 211

smart materials, 1, 232

socket, 42, 127, 142

solution, 166, 184

speed, 44, 142

standing phase, 36, 38, 52, 53

status, 50, 169

steadystate, 120

storage, 45, 85, 125, 140, 183, 184, 199

storing, 52, 79

structure, 8, 10, 11, 17, 19, 20, 22, 24, 86, 89, 91, 101, 129, 130, 133, 188, 213 stump, 52, 125, 141

substantially, 26

substrate, 7, 17, 23, 88,
   130, 132, 133

suitable, 55, 165, 168,
   170, 180, 183, 231,
   238

Surface, 21, 24

surplus, 84

swing-phase, 36

switched off, 209

switching, 4, 167, 168,
   169, 185, 237

techniques, 27, 90, 99

technology, 4, 15, 33,
   66, 237

temperature, 25, 26

terminal, 107

test, 44, 50, 51, 55, 84,
   87, 209

test kit, 44

testing, 196, 199, 211,
   218, 219, 223

thermal, 25, 66

third position, 75

time, 3, 31, 44, 142,
   151, 200, 209, 222,
   236

transducer, 32

transform, 66, 95

transmission, 85

transmitter, 31

ultra low power, 187

underdamped, 105

underneath, 62, 70, 73

unregulated, 116

upper, 1, 41, 42, 68, 77, 131, 136, 137, 138, 232

utilization, 83, 129

utilized, 83, 121, 123, 235

value, 98, 102, 142, 152, 192, 194, 195, 202, 205, 208, 210, 223, 224, 225, 227, 228, 229, 230, 238

variation, 38, 44, 53, 78

verified, 83, 123, 234

vertical, 69

vibration, 44, 66, 82, 84, 92, 101, 126, 185, 211, 214, 215

vibrations, 3, 186, 215, 236

view, 37, 39, 41

voltage, 5, 9, 24, 26, 67, 71, 72, 74, 75, 76, 77, 87, 98, 99, 100, 101, 103, 104, 105, 108, 113, 114, 115, 118, 119, 120, 126, 129, 134, 135, 137, 138, 139, 140, 144, 145, 148, 149, 151, 152, 153, 154, 155, 156, 157, 158, 162, 164, 165, 166, 167, 169, 171, 172, 173, 174, 176, 178, 190, 192, 195, 202, 203, 204, 205, 207, 208, 209, 218, 220, 221, 223, 225, 229, 230, 237

voltage source, 135,
222

Vpulse, 108

walking, 3, 28, 48, 50,
51, 55, 56, 58, 59, 77,
125, 141, 155, 177,
236

waveform, 73, 98, 100,
110, 111, 148, 149,
153, 157, 205, 220,
221

waveforms, 100, 129,
136, 137, 138, 139,
140, 145, 154, 155,
156, 157, 158, 205

wears, 125, 141

wireless, 187

Piezoelectrics in Prosthetics: Energy Harvesting

PECHRACH PUBLISHING

You may be interested in other titles from Pechrach publishing:

PIEZOELECTRICS
IN
CIRCUIT BREAKERS
Design & Test

Paperback: 180 pages
Language: English
ISBN-10: 0993117805
ISBN-13: 978-0993117800
Product Dimensions: 14 x 1 x 21.6 cm

This book demonstrates how a piezoelectric actuator can be used as part of the actuation system in a circuit breaker mechanism. The design of circuit breaker systems could be simplified by the use of smart materials. The simpler the operating mechanism becomes, the number of mechanical parts can be minimized. This will improve reliability, reduce size, power consumption and manufacturing costs. It allows substantial miniaturization of these devices resulting in a completely new type of circuit breakers.

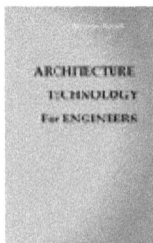

Paperback: 268 pages
Language: English
ISBN-10: 099311783X
ISBN-13: 978-0993117831
Product Dimensions: 15.2 x 1.7 x 22.9 cm

This Architecture Technology for Engineers book is a basic principles for first year architecture students. The benefit to engineers, this book will help to have a better understanding the reason behind the architecture ideas and design. This book consists of 12 chapters cover overall knowledge of architecture structure such as monolithic, frame and shell, how to conduct a site surveys, sketching, drawings, and diagram, learn about materials, sources of materials, the sequence of construction which includes foundation, floor, walls and roof, the affect of the environment, solar energy, thermal control, heat gain and heat loss, the carbon emissions, climate analysis, health and well- being in architecture, the services includes domestic services, fresh water, waste water, hot water, electricity, energy saving, heating and ventilation and how to develop and modify the building to be sustainability. There are exercises and case studies at the end of each chapter.

Paperback: 260 pages
Language: English, Thai
ISBN-10: 0993117813
ISBN-13: 978-0993117817
Product Dimensions: 13.3 x 1.7 x 20.3 cm

This book is written in two languages: Thai and English. You can create the way you want your life to go. If you really want something, you can find a way. You have to know who you are, what your advantages and disadvantages are. When you tell yourself how good you are, you feel good. You see pictures of yourself confident and successful. When you change your way of thinking, you will feel your emotions changing. Wisdom creates confidence. Charm can be created by practising meditation. When your mind is peaceful, you can see how to solve problems and how you deal with them. You can find money whenever you need it. You create your income from inspiration, creativity and skills.

Paperback: 464 pages

Language: English

ISBN-10: 0993117864

ISBN-13: 978-0993117862

Product Dimensions: 13.3 x 2.7 x 20.3 cm

Slow Contact Opening
Circuit Breakers

New results in this book show that at the point at which the arc root moves from the contact region, the driven flow could dominate the arc root commutation from the contact region. The arc root contact time analysis and modelling of the magnetic forces lead to able to estimate the minimum forces that moved the arc root from the contact region. The minimum force for the arc to move from the contact region, with contact opening velocities between 1 m/s and 10 m/s, was approximately 0.2 N. The short circuit current showed a significant influence on the movement of the arc root and the magnetic forces on the contact area and the arc root contact time is decreased as the contact gap increases for contact velocities below 2.2 m/s. The contact opening velocity has a significant influence on the mobility of the arc root from the contact region. The arc root motion depended upon the quantity of the flow and area of the arc chamber venting.

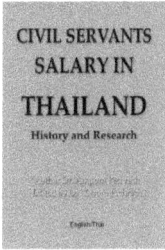

Paperback: 552 pages

Language: English

ISBN-10: 0993117856

ISBN-13: 978-0993117855

Product Dimensions: 13.3 x 3.2 x 20.3 cm

CIVIL SERVANTS SALARY IN THAILAND
History and Research
English/Thai

This book is written in English and Thai languages. The researcher constructs a base-salary model and a corresponding base-salary scale table for government officers that are adjustable to changing economic conditions. This book adopts an econometric approach in accordance with which stochastic frontier analysis (SFA) was conducted by means of an application of the maximum likelihood estimation (MLE) technique. The base - salary of the military officers is adjustable to prevailing economic conditions. The objective is to be able to calculate the maximum/minimum rate of the salary structure. It also aims to calculate the step-range for each work group.

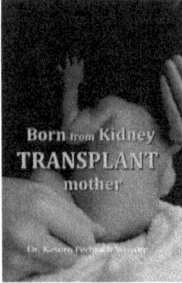

Paperback: 552 pages

Language: English

ISBN-10: 0993117856

ISBN-13: 978-0993117855

Product Dimensions: 13.3 x 3.2 x 20.3 cm

This Born from Kidney Transplant Mother book is written from the author's experience as kidney transplant. She was in the end stage kidney failure and received a live kidney transplant from her sister. Afterward, she had gone to have three experienced pregnancy with miscarriage, stillbirth and prematurity. It is based on some theory and 12 years of data collections about the condition of the kidney transplant mother before pregnancy, during pregnancy and after giving birth including the development and growth of her premature baby. This book would be useful for the kidney transplant woman who wants to have a baby, medical staffs and people who interest in kidney transplant.

www.ingramcontent.com/pod-product-compliance
Lightning Source LLC
Chambersburg PA
CBHW020833210326
41598CB00019B/1881